W0036971

Springer Series in Computational Physics

Editors

W. Beiglböck   H. Cabannes
H. B. Keller   J. Killeen   S. A. Orszag

Finite-Difference Techniques for Vectorized Fluid Dynamics Calculations

Edited by David L. Book

With Contributions by

Jay P. Boris
Martin J. Fritts
Rangaro V. Madala
B. Edward McDonald
Niels K. Winsor
Steven T. Zalesak

With 60 Illustrations

Springer-Verlag
New York Heidelberg Berlin

David L. Book
Naval Research Laboratory
Washington, DC 20375
USA

Editors

W. Beiglböck
Institut für Angewandte Mathematik
Universität Heidelberg
D-6900 Heidelberg 1
Federal Republic of Germany

H. Cabannes
Mécanique Théoretique
Universite Pierre et Marie Curie
F-75005 Paris
France

H. B. Keller
California Institute of Technology
Pasadena, CA 91125
USA

J. Killeen
Lawrence Livermore Laboratory
Livermore, CA 94551
USA

S. A. Orszag
Massachusetts Institute of Technology
Cambridge, MA 02139
USA

Library of Congress Cataloging in Publication Data

Main entry under title:

Finite-difference techniques for vectorized fluid
   dynamics calculations.
   (Springer series in computational physics)
   Bibliography: p.
   Includes index.
   1. Fluid dynamics. 2. Finite differences.
I. Book, D. L. (David Lincoln), 1939–
II. Series.
TA357.F49   532'.05'0151562   81-5282
                                    AACR2

ISBN-13: 978-3-642-86717-0   e-ISBN-13: 978-3-642-86715-6
DOI: 10.1007/ 978-3-642-86715-6

Summary

This book describes several finite-difference techniques developed recently for the numerical solution of fluid equations. Both convective (hyperbolic) equations and elliptic equations (of Poisson's type) are discussed. The emphasis is on methods developed and in use at the Naval Research Laboratory, although brief descriptions of competitive and kindred techniques are included as background material. This book is intended for specialists in computational fluid dynamics and related subjects. It includes examples, applications and source listings of program modules in Fortran embodying the methods.

Contents

David L. Book, Jay P. Boris, and Martin J. Fritts are from the Laboratory for Computational Physics, Naval Research Laboratory, Washington, D.C.
B. Edward McDonald, Rangaro V. Madala, Niels K. Winsor, and Steven T. Zalesak are from the Plasma Physics Division, Naval Research Laboratory, Washington, D.C.

CHAPTER 1

Introduction

In this book we describe a series of vectorizable computational techniques for solving hydrodynamics problems which have been developed at the Naval Research Laboratory in the past few years. We also relate them to older techniques, discuss the applications for which they were developed, and describe their implementation on a particular vector machine, the NRL Texas Instruments Advanced Scientific Computer (ASC).

A tremendous range of physical phenomena are described by fluid equations. All four of the classical "elements"—earth, water, air, and fire—behave, at least on some scales, as fluids. For this to be true, the features they share in common must be very simple, in spite of the variegation and superficial complexity we are accustomed to dealing with. This is indeed the case, and the present book deals chiefly with those central aspects.

The mathematical description of fluid motion makes use of partial differential equations which propagate the fluid variables (density, velocity, pressure, etc.) in time. Most systems of such equations closely resemble one another in containing convective derivatives, expressing causality, conservation laws, and other properties, and in emphasizing the connectedness of the physical system. This latter property is often manifested in the appearance of a Poisson equation, which relates the pressure or velocity potential at a given point to the properties of the physical system as a whole. It is also displayed in the smooth (rather than discontinuous) variation through most of the system which we expect in fluids. However, many of the most important computational methods are intended to describe discontinuous behavior as well, such as that occurring in shocks, at combustion fronts, and on free surfaces and boundaries.

Because we have emphasized the common features and generality of fluid behavior, most of the discussion in this book is theoretical, and the applications given are fairly simple. To apply the techniques to specific problems of interest, one or more of the following must be added to the basic (Navier–Stokes) fluid equations: (i) more dimensions; (ii) more complicated geometry or boundary conditions; (iii) additional source, sink, or transport terms; (iv) more dependent variables; (v) an equation of state. For example, ideal magnetohydrodynamics requires the introduction of the magnetic field strength $\mathbf{B}$ and the Lorentz force; nonideal magnetohydrodynamics requires thermal conduction, resistivity, and viscosity terms. Ocean internal waves require a buoyancy term and an eddy viscosity derived from a turbulence closure model. Chemically reactive flow requires the introduction of chemical variables, together with reaction rates, thermal conduction, and a turbulence model. Although most of the *physics* in any given problem is comprised in these added terms and conditions, computational practice shows that most of the difficult *numerics* problems are associated with the universal fluid features. If adequate care is devoted to developing accurate general numerical techniques for treating these, the resulting code invariably turns out to be robust, adaptable, and efficient. Hence the techniques described here encompass a substantial portion of the effort needed to construct even the most elaborate numerical hydrodynamic model.

We discuss techniques for both "incompressible" and "compressible" hydrodynamics, distinguished according to whether $\nabla \cdot \mathbf{v} = 0$ or $\nabla \cdot \mathbf{v} \neq 0$. In general we will be treating systems in which the fundamental variables are the mass density ρ, flow velocity $\mathbf{v}$, and pressure p, and the equations are, respectively, those of continuity or advection; momentum transfer (Euler's equation); and an energy transport equation, equation of state, or the incompressibility condition (see Section 2.1).

It is assumed that the reader is familiar with the basic concepts of fluid dynamics, as expounded, for example, in Landau and Lifshitz (1959). No attempt will be made in this book to study in any detail the physics of fluids, or to explore the physical significance of the specialized problems which have been solved using the computational methods we present. Instead the emphasis is on the methods themselves.

The reader is expected to be conversant in the standard numerical techniques used for machine computation, as given in, e.g., Hamming (1962), and to have had experience in programming physics problems in Fortran or some other high-level language. Moreover, he should have some acquaintance with standard methods for computer solution of fluid problems, such as would result from having been involved in research in which fluid computations were carried out or from having read the descriptions given by, e.g., Richtmyer and Morton (1967) or Roache (1975). This book is intended, in other words, for the specialist in computational fluid dynamics or in one of the fields which makes use of fluid computational techniques: plasma, space, solar or astrophysics, oceanology, meteorology, combustion, etc.

As used in this book, "computational techniques" relate to the solution of equations derived from theoretical models, specifically models of hydrodynamic phenomena, and are to be distinguished from the use of computers in data reduction and analysis. Almost all of the techniques are aimed at solving nonlinear problems, which are inherently intractable by classical analytic techniques. Indeed, it is to their nonlinearity that fluid dynamic systems owe both their richness and their relative inaccessibility to computational treatment.

Every theoretical model yields its predictions as solutions of the model equations. These may be as general as the three-dimensional Navier–Stokes equations, or as specific as an equation for the evolution of a particular type of wave (e.g., solitons) or a steady-state turbulent spectrum. Broadly speaking, there is a correlation between generality and the following difficulties:

1. Heavy machine requirements (storage and CP time);
2. need for highly sophisticated numerical methods;
3. limitations in precision and validity of long-term predictions;

while specificity is associated with

1. theoretical problems in deriving and justifying the model;
2. stringent requirements on plausibility and agreement of predictions with experiment.

In addition to determining the concrete predictions inherent in a particular set of assumptions, computer modeling can be used in an exploratory role, through "computer experiments." Where the precise set of assumptions or the values of the physical parameters are not clear, it is possible to generate a family of solutions to study trends and systematic properties. This form of computation is often referred to as "numerical simulation" and is contrasted with "numerical solution" of equations.

This computer-experimentation point of view is the one adopted in the present work. We will be concerned principally with addressing the three kinds of difficulty listed above in connection with general models. The techniques we describe are designed to work in a production mode which emphasizes convenient data structures, high resolution (large computational meshes), detailed diagnostics, ruggedness, and, so far as is compatible with these, high execution rates.

Most techniques which are inherently efficient show to good advantage on any machine and can be coded in nearly machine-independent form. In order to shave the last few percent (sometimes more than a few) from execution times, it is essential to make optimum use of the characteristics of the individual machine. The methods described in the text and the corresponding modules listed in the Appendices were developed for the NRL ASC. While some have been run on other machines with only small changes, optimization would entail a certain amount of recoding. The effort required varies, depending on both the machine and the routine involved.

This book is organized as follows. In Chapter 2 six different classes of methods are compared and categorized in a framework provided by a set of six requirements which ought to be satisfied by an ideal continuity equation algorithm. The strengths and weaknesses of the different approaches are analyzed. Within this framework finite-difference techniques have particularly attractive properties. Hence most of the book is devoted to describing techniques which correct, at least partially, the difficulties from which finite-difference techniques suffer. Chapter 3 is devoted to flux-corrected transport (FCT) and related Eulerian solutions of the nonlinear continuity equation, which preserve the positivity property of the real equations and yet have relatively low numerical diffusion. Chapter 4 describes another kind of Eulerian code, designed to solve evolution equations describing the upper ocean and the atmosphere without the stringent time-step limitation imposed by an explicit treatment of surface waves and fast-moving internal gravity waves. Chapter 5 describes ADINC, a Lagrangian algorithm appropriate for describing nearly incompressible flow. In Chapter 6 we consider the Lagrangian techniques incorporated in the SPLISH code, which uses a reconnectable grid of triangles to allow accurate numerical solution of complicated flows with breaking waves and other severe nonlinearities. In Chapter 7 we briefly survey elliptic equation solvers, then describe two new techniques—one direct and one iterative—which have proven useful in fluid computations. Chapter 8 describes the rationale behind vectorization and summarizes techniques used to vectorize codes we have written incorporating the techniques described in this book. From several of these codes we have extracted the core modules, Fortran listings of which are collected in the Appendices, along with directions for utilization.

While this book is in no sense a treatise on the whole subject of computational fluid dynamics, it spans a wide range of topics. These were selected because each represents a definite advance in the state of the computational art. Although they have been described in specialist journals, each of the techniques is useful in a wide variety of applications. By collecting them in this fashion, relating them to each other and to previously existing techniques, and describing both theory and applications, we hope to bring about their exposure to a much wider audience than would otherwise be possible.

CHAPTER 2

Computational Techniques for Solution of Convective Equations

Here we survey the computational techniques appropriate to fluid motion, with emphasis on those of greatest utility and already implemented, along with remarks about hardware and utility software requirements. In several cases a technique may have application to physical problems other than those of fluid dynamics. No attempt has been made at completeness. Rather, we mention those techniques which are most widely used, most typical of a class or family, best suited to introduce the reader to a key concept, or most closely related to the new methods we are describing. Likewise, the references are for the most part to well-known papers and monographs, research we consider of particular importance in relation to our own, or simply the relevant publications with which we are most familiar.

2.1 Importance of Convective Equations

For a two- or three-dimensional incompressible flow, one can either choose to treat the primitive equations

$$\frac{\partial \rho}{\partial t} + \mathbf{v} \cdot \nabla \rho = 0, \tag{2.1}$$

$$\nabla \cdot \mathbf{v} = 0, \tag{2.2}$$

$$\rho \left(\frac{\partial \mathbf{v}}{\partial t} + \mathbf{v} \cdot \nabla \mathbf{v} \right) + \nabla p = 0, \tag{2.3}$$

or one can replace (2.2) and (2.3) with the equivalent vortex dynamics equations

$$\mathbf{v} = \nabla \times \mathbf{A}, \tag{2.4}$$

$$\nabla^2 \mathbf{A} = -\zeta, \tag{2.5}$$

$$\frac{\partial \zeta}{\partial t} + \nabla \cdot (\mathbf{v}\zeta) = \zeta \cdot \nabla \mathbf{v} + \rho^{-2} \nabla \rho \times \nabla p, \tag{2.6}$$

where $\mathbf{A}$ is the vector stream function and ζ is the vorticity. In two dimensions the stream function and vorticity are scalars and (2.4)–(2.6) become

$$\mathbf{v} = \hat{\mathbf{z}} \times \nabla\psi, \tag{2.7}$$

$$\nabla^2\psi = \zeta, \tag{2.8}$$

$$\frac{\partial\zeta}{\partial t} + \nabla\cdot(\mathbf{v}\zeta) = \rho^{-2}\hat{\mathbf{z}}\cdot\nabla\rho \times \nabla p. \tag{2.9}$$

The pressure is found, if required, by solving the elliptic equation obtained from taking the divergence of (2.3) and using (2.2):

$$\nabla^2 p = \frac{1}{\rho}\nabla\rho\cdot\nabla p - \rho\nabla\mathbf{v}:\nabla\mathbf{v}. \tag{2.10}$$

Either formulation involves convective equations [i.e., (2.1), (2.3), (2.6), (2.9)], which therefore assume a central role in numerical fluid dynamics. In addition, the vorticity formulation involves equations of Poisson's type, (2.5) or (2.8), as well as (2.10). In this section we treat only the former; the latter are discussed in Chapter 7. We ignore transport terms (heat conduction, molecular or eddy viscosity, diffusion) which are usually small except in very narrow regions, and in any case are computed as corrections that do not affect the convective properties. Likewise, although boundary conditions often play a crucial role, we mention them for the most part only in passing. This is because the standard techniques for handling boundaries are well developed and well described in, e.g., Roache (1972), and apply to most of the methods we are describing.

For compressible flows, (2.1) and (2.2) are replaced by

$$\frac{\partial\rho}{\partial t} + \mathbf{v}\cdot\nabla\rho + \rho\nabla\cdot\mathbf{v} = 0 \tag{2.11}$$

and an equation of state in differential form, e.g., for an ideal gas with adiabatic index γ,

$$\frac{\partial p}{\partial t} + \mathbf{v}\cdot\nabla p + \gamma p\nabla\cdot\mathbf{v} = 0, \tag{2.12}$$

or an equation for the energy density $\mathscr{E}$,

$$\frac{\partial\mathscr{E}}{\partial t} + \nabla\cdot[\mathbf{v}(p + \mathscr{E})] = 0, \tag{2.13}$$

where for an ideal gas

$$\mathscr{E} = \frac{1}{2}\rho v^2 + \frac{p}{\gamma - 1}. \tag{2.14}$$

When $\nabla\cdot\mathbf{v} \neq 0$, $\mathbf{v}$ must be derived in general from both a vector and a scalar potential:

$$\mathbf{v} = \nabla \times \mathbf{A} + \nabla\phi. \tag{2.15}$$

Both **A** and ϕ satisfy Poisson equations. The source of the latter,

$$\sigma = \nabla \cdot \mathbf{v}, \tag{2.16}$$

satisfies the convective equation which results from taking the divergence of (2.3):

$$\frac{\partial \sigma}{\partial t} + \nabla \cdot (\mathbf{v}\sigma) = \sigma^2 - \nabla \mathbf{v} : \nabla \mathbf{v} + \rho^{-2}(\nabla \rho \cdot \nabla p - \rho \nabla^2 p). \tag{2.17}$$

Generally speaking, no advantage results from using the formulation in terms of ζ and σ in problems of compressible flow.

Both (2.12) and (2.13) are convective equations and can be solved by the same methods. The distinction between the two is important, however, as the former permits only adiabatic (unshocked) flows. The only difficulty with using (2.13) is the necessity of extracting p from (2.14); when $p \ll \mathscr{E}$, truncation errors often result in negative pressures, which are then usually set equal to zero. Some techniques avoid this difficulty entirely, as we will see shortly, but it is routinely present in finite-difference treatments.

2.2 Requirements for Convective Equation Algorithms

The following sections review the different numerical techniques which have been developed to solve fluid dynamics problems based on the advective equation (2.1) or the continuity equation (2.11) (written in conservation form in terms of a general variable ρ),

$$\frac{\partial \rho}{\partial t} + \nabla \cdot \rho \mathbf{v} = 0. \tag{2.18}$$

Equation (2.18) is a statement about the properties of a continuum. That is, ρ is an infinite-dimensional vector which may be variously represented in configuration space [as $\rho(\mathbf{x})$], in Fourier space or some other spectral decomposition with a countably infinite set of coefficients, in an infinite Taylor series, as the limit of a superposition of step functions, etc. Numerical approximations to ρ can be thought of as projections of this vector onto a finite-dimensional subspace. When we approximate the evolution Eq. (2.18) numerically, we attempt to devise a means of predicting, on the basis of a finite approximation at some time t (and possibly at several other neighboring time levels $t - \delta t$, $t - 2\delta t$, etc.), what the equivalent finite approximation to ρ will be at time $t' > t$. It is easy to show, by a host of counterexamples, that this simply cannot be done in general. There is an "uncertainty principle" for all numerical approximations to a given linear continuum evolution equation which says, roughly speaking, that the more erratically the original continuum description ρ varies, the less reliably its discrete representation can be extrapolated in time.

For nonlinear systems the uncertainty is correspondingly greater. Consequently, the most we can ask of a convective equation solver is that it generate solutions which reliably approximate the temporal evolution of some subset of the initial continuum state vectors belonging to the particular initial discrete representation. The essence of the art of computation is in defining algorithms which closely track the evolution of those continuum states which are physically significant.

Equation (2.18) is written in conservation form. There are other ways to write the continuity equation, which are reflected in some of the numerical solution techniques:

$$\text{Convection form}: \frac{\partial \rho}{\partial t} + \underbrace{\mathbf{v} \cdot \nabla \rho}_{\text{convection}} = \frac{d\rho}{dt} = - \underbrace{\rho \nabla \cdot \mathbf{v}}_{\text{compression}} ; \tag{2.19}$$

$$\text{Integral form}: \frac{\partial}{\partial t} \int_{\text{region}} \rho d^3\mathbf{r} = - \int_{\text{boundary}} \varrho \mathbf{v} \cdot d\mathbf{A}. \tag{2.20}$$

The convective term $\mathbf{v} \cdot \nabla \rho$ displayed explicitly in (2.19) gives it its intrinsically hyperbolic form and causes severe problems numerically. The compression term $- \rho \nabla \cdot \mathbf{v}$ is absent in incompressible hydrodynamics applications. Version (2.1) without this term is often called the convection (or advection) equation.

Convective equations appear in almost all dynamic descriptions of physical continua simply because they express two of the general principles in physics, conservation and causality. The integral form given by (2.20) states that the amount of material $\int \rho d^3\mathbf{r}$ in a given region can only change due to flows of material $\rho \mathbf{v} \cdot d\mathbf{A}$ through the boundary (conservation). Any material currently at a given point could only have gotten there by previously leaving someplace else (causality) and passing through the points between. Continuity equations underlie compressible and incompressible fluid dynamics, hydrodynamics, plasma physics, and quantum mechanics.

The convective Eq. (2.1) or (2.11) displays the positivity property. That is, if $\rho \geq 0$ everywhere in the system at some instant in time, then this property must hold at all later times also, provided ρ is analytic. For otherwise at some time t_0 there would be some point $\mathbf{x}_0$ at which ρ first becomes negative. There ρ satisfies

$$\rho(\mathbf{x}_0, t_0) = 0 = \nabla \rho(\mathbf{x}_0, t_0) \tag{2.21}$$

since $\mathbf{x}_0$ is a minimum. Then from (2.11),

$$\frac{\partial \rho}{\partial t} = - \rho \nabla \cdot \mathbf{v} - \mathbf{v} \cdot \nabla \rho = 0. \tag{2.22}$$

But

$$\frac{\partial^2 \rho}{\partial t^2} = - \frac{\partial \rho}{\partial t} \nabla \cdot \mathbf{v} - \rho \nabla \cdot \frac{\partial \mathbf{v}}{\partial t} - \mathbf{v} \cdot \nabla \frac{\partial \rho}{\partial t} - \frac{\partial \mathbf{v}}{\partial t} \cdot \nabla \rho = 0 \tag{2.23}$$

also, and similarly for all higher derivatives. Accordingly, for $t = t_0 + \tau$,

$$\rho(\mathbf{x}_0, t) = \rho(\mathbf{x}_0, t_0) + \tau \frac{\partial \rho}{\partial t}(\mathbf{x}_0, t_0) + \frac{1}{2}\tau^2 \frac{\partial^2 \rho}{\partial t^2}(\mathbf{x}_0, t_0) + \cdots = 0, \quad (2.24)$$

contradicting the assumption. Thus a quantity being transported will never turn negative anywhere if it was initially everywhere positive. Material cannot be removed from a region which is already devoid of material. The fact that matter obeys particle-like and fluid-like equations on microscopic and macroscopic scales, respectively, has its counterpart for numerical solution techniques. The first three of the techniques described below model the system as an aggregate of discrete entities, while the last three aim more directly at solving the partial differential equation by discretizing the equation itself.

The convective equation is often termed hyperbolic, even though it has only first derivatives in space and time, because it permits propagating (wave) solutions. It is quite different from the second-order wave equation, which is normally called hyperbolic, and which is equivalent to a system of two coupled first-order convective equations. Therefore, the solution techniques can differ significantly. Generally the characteristics of the equation are in one direction only. Wave equations usually occur in pairs with two oppositely directed characteristics, and each of the two quantities being propagated (say $\mathbf{E}$ and $\mathbf{B}$ in electromagnetics) appears in the time derivative of one equation and in the space derivative of the other. Thus second-order wave equations have no positivity to speak of, and the most useful conservation property is quadratic rather than linear. Because of these distinctions, some of what is said here about convective equations does not apply to second-order wave equations.

This brief survey considers only conservative solution techniques, although nonconservative techniques are adequate for some problems. In most situations the conservation properties of the physics have to be mirrored in the numerical method, or else nonphysical instabilities or unacceptable secular errors will result. Although positivity is frequently as important as conservation, not all of the methods generally applied have reflected this fact. The most common and annoying problems which arise in the solution of coupled systems of convective equations occur because the quantity convected becomes falsely negative somewhere. This is related to the occurrence of numerical oscillations in the vicinity of steep gradients and discontinuities.

The analysis here is limited to the basic methods, although there are a host of hybrid techniques combining two or more of the basic approaches. The number of such combinations is staggering, and even the definitions of the basic methods are in some dispute. Therefore no useful purpose is served by trying to classify and analyze all of the possible combinations here. Rather, the general properties of each are described, and the pros and cons considered.

As a framework for evaluating the algorithms, we have selected six general requirements which an ideal numerical algorithm for solving the convective equation should satisfy to be generally useful. These requirements are given in what we estimate to be their order of importance. An ideal algorithm should

1. be linearly stable for all cases of interest;
2. mirror conservation properties of the physics;
3. ensure the positivity property when appropriate;
4. be reasonably accurate;
5. be computationally efficient;
6. be independent of specific properties of one particular application.

This last requirement restates our preference for useful general methods. As with the methods themselves, some of the requirements are related. Stability, conservation, and positivity generally relate to accuracy, for example, and yet no one of these can guarantee any of the others. On the other hand, the requirements can also be partially contradictory. Accuracy and efficiency, for example, often tug in opposite directions.

The following list shows the six methods to be discussed, the order being chosen to range from most discrete (or particulate) in nature to most continuous (or global).

- Quasiparticle Methods
- Characteristic Methods
- Lagrangian Finite-Difference Methods
- Eulerian Finite-Difference Methods
- Finite-Element Methods
- Spectral Methods

2.3 Quasiparticle Methods

In their simplest form quasiparticle methods attempt to solve (2.1)–(2.3) by simulating the microscopic physics of the particles which make up the fluid. If a large number N of quasiparticles are initialized in the calculation at positions $\{X_j(0)\}$ at time $t = 0$, their later positions can be found at time t by integrating the simple orbit equation

$$\frac{dX_j}{dt} = V_j(t), \tag{2.25}$$

where the velocity V_j is found by integrating a force law

$$\frac{dV_j}{dt} = F_j(x, V, t). \tag{2.26}$$

Here the force F_j can depend on the positions and velocities of particles other than the jth particle. Then the fluid mass density is found by summing the masses m_j of all the particles j at or near the ith grid point,

$$\rho(x_i) = \left\langle \sum_j m_j \delta(X_j - x_i) \right\rangle, \tag{2.27}$$

the fluid momentum density is found by summing the particle momenta,

$$\rho\mathbf{v}(x_i) = \left\langle \sum_j m_j \mathbf{V}_j \delta(\mathbf{X}_j - \mathbf{x}_i) \right\rangle, \tag{2.28}$$

etc. Here the angle brackets refer to the type of averaging used to smooth out the discreteness, generally some form of area-weighting (linear interpolation) of finite-sized particles among several cells.

Equations (2.25) and (2.26) are ordinary differential equations and are usually solved easily with a high degree of accuracy. When each quasiparticle has a velocity which is determined solely as a function of the particle position at any time, i.e.,

$$\mathbf{V}_j(t) = \mathbf{V}[\mathbf{X}_j(t), t], \tag{2.29}$$

the quasiparticles cannot pass each other, except as a result of finite time-step errors [or perhaps singularities in $\mathbf{V}(\mathbf{X}, t)$]. The model is then appropriate for collisional fluids. Nonetheless, the incompressibility condition, $\nabla \cdot \mathbf{v} = 0$, is very difficult to enforce self-consistently and rigorously in these models. When several quasiparticles can occupy essentially the same location in space and yet be moving with significantly different velocities, the model is more appropriately used to describe "collisionless" systems, such as plasmas on the microscopic scale or, on a grander scale, stars in self-gravitating systems.

The term quasiparticle has been used because it is almost never possible to carry in the computer memory as many particles as are involved in the actual physical problem of interest. It is not unusual to simulate systems of from 10^{11} particles (galaxies) to 3×10^{22} particles (1 cm^3 of water) by using 10^3–10^6 quasiparticles. In fact, when the number of quasiparticles exceeds 10^3 or so, it is impractical to treat the forces by calculating a direct interaction law. Then a potential ϕ is introduced according to

$$\mathbf{F}_j = -\nabla\phi(\mathbf{X}_j). \tag{2.30}$$

This potential satisfies a Poisson equation of the form (2.8) relating the source terms (e.g., charge or mass) and the corresponding long-range fields ϕ. It is solved on a finite-difference grid in either configuration or transform space (see Chapter 7).

The collisional quasiparticle methods which have been tried to date include PIC (Evans and Harlow, 1957), MAC (Harlow and Welch, 1965), GAP (Marder, 1975), VORTEX (Christiansen, 1973) and more recent attempts of the same genre by workers at Los Alamos (Taggart et al., 1975). These methods introduce what amounts to a finite-difference grid for some of the derived physical quantities, such as pressure, while retaining the particle mass, momentum, and energy as variables. The collisionless quasiparticle methods are reviewed by Birdsall and Fuss (1969), by Langdon and Lasinski (1976), and in the *Proceedings of the Fourth Conference on Numerical Simulation of Plasma*, edited by Boris and Shanny (1970). The original applications seem to stem from work by Hockney (1965, 1966) and Buneman (1967).

Typical of the elaborations which have grown out of the simplistic "particle-pushing" approach outlined above is the Particle-in-Cell (PIC) method (Harlow, 1964). It employs a rectilinear Eulerian mesh, within each cell of which are defined a mean velocity, density, internal energy, and pressure. In the first stage of the calculation the particles are regarded as fixed. The density ρ_α is taken to be the mass m_α times the number of particles of species α within the given cell, divided by the volume associated with the cell. The pressure is calculated from the partial pressures p_α, derived from ρ_α using the respective equations of state.

Next, (2.3) and (2.13) or an analogous equation for the internal energy density $U = \mathscr{E} - \frac{1}{2}\rho v^2$ are solved without the advection terms to obtain intermediate values of v and $\mathscr{E}$ or U. The equations are differenced so as to conserve momentum and energy exactly, the latter by using the mean of the old and the intermediate velocities in the compression term. The particles are then translated using the area-weighted intermediate velocities. Some of them cross cell boundaries. In so doing they transfer mass, momentum, and energy from the old cell to the new one, causing the Eulerian velocities, energy densities, and mass densities to be updated accordingly.

The PIC method shares with other particle-pushing techniques the problem of graininess when the number of particles of a given species per cell becomes small. In addition, the Eulerian differencing scheme used is unstable, and the method succeeds only because of the effective viscosity $\bar{\mu} \sim \frac{1}{2}\bar{\rho}\bar{v}\delta x$ associated with particles crossing grid boundaries. Near stagnation regions where $v \rightarrow 0$ the results are therefore inaccurate. There are also resolution problems associated with extreme expansions and weak shocks, where an artificial viscosity (Richtmyer and Morton, 1967) is needed.

The Marker-and-Cell (MAC) method (Harlow and Welch, 1965) is essentially an Eulerian finite-difference scheme (see Section 2.5) in which marker particles are advected passively by the computed flow field. The markers participate in the dynamics only in determining the position of a free surface. This is important, however, owing to the extensive use of MAC in surface wave problems.

The Grid-and-Particle (GAP) method of Marder (1975), on the other hand, differs from PIC in that the particles carry around all fluid properties, including specific volume. Area weighting is used to determine p and ρ on the grid and then to calculate from them the forces pushing particles. GAP treats shocks of any strength without need of artificial viscosity and has somewhat better resolution and stability properties than PIC.

The quasiparticle methods are generally stable in practice and conserve where they should (almost by definition). These methods can also guarantee positivity since it is just as hard to remove a quasiparticle from an already empty region of space as it is a real particle. The methods can even be made reasonably general and flexible. Where they fail is in efficiency and accuracy. Because the statistics and hence the smoothness of the computed solutions depend on the number of quasiparticles which the user can afford, many applications of interest are beyond the particle approach. Three-dimensional problems and long-time turbulent incompressible flows are unacceptably expensive.

2.4 Characteristic Methods

Ideal one-dimensional plane (i.e., rectilinear) isentropic compressible flow can be described in terms of the Riemann invariants

$$R_\pm = v \pm \int \frac{dp}{\rho c}, \tag{2.31}$$

where $c = (\partial p/\partial \rho)^{1/2}$ is the speed of sound. The quantities $R_\pm$ are constant along trajectories $C_\pm$, defined by

$$\frac{dx}{dt} = v \pm c, \tag{2.32}$$

respectively (Landau and Lifshitz, 1959; Courant and Friedrichs, 1948). If discontinuties or increases occur in the entropy s, they are propagated according to the adiabatic equation

$$\frac{\partial s}{\partial t} + v\frac{\partial s}{\partial x} = 0, \tag{2.33}$$

i.e., s is constant along trajectories C_0 associated with the flow speed,

$$\frac{dx}{dt} = v. \tag{2.34}$$

Equations (2.31)–(2.34) completely describe flows in which $s = $ const everywhere in terms of motions along the characteristic trajectories, or characteristics, denoted by $C_\pm$ and C_0. An arbitrary small perturbation can also be written as a superposition of disturbances propagating according to these three laws.

In anisentropic flows, two types of discontinuities (interfaces between different states of the same medium) can occur. These are contact discontinuities, across which density, energy, and entropy are discontinuous, but pressure and normal velocity (in one dimension, the only component) are continuous, and shocks, across which all dependent variables change discontinuously. Contact discontinuities propagate with the flow speed, i.e., along C_0. Shock waves move with a velocity V which uniquely determines (through the Rankine–Hugoniot relations) the discontinuous changes in ρ, v, and p. In the limit of very weak shocks, the shock trajectory reduces to one of $C_\pm$, i.e., to that of an acoustic disturbance.

By modeling an arbitrary flow as a series of isentropic regions separated by discontinuities, it is possible to develop a general computational technique. In this approach it is necessary to distinguish six different kinds of points (Hoskin, 1964): those interior to isentropic regions, system boundaries, free surfaces, contact discontinuities, single shocks, and intersecting shocks. These points are labeled with the coordinates x, t; in general, the latter varies from point to point, so we do not work on a single "time level." To calculate the fluid properties at a new point on the intersection of the characteristics running through two known points, however, only the physical quantities at those points are needed. Thus it is necessary to store only an amount of information equivalent to that on one time level of a finite-difference scheme. In fact, some characteristic schemes use interpolation onto a grid of points all at the same time t, permitting the calculation to "march" from level to level.

Characteristic (also known as shock-fitting) methods which take advantage of the physical content of the specific system of equations being solved are not entirely counter to the sixth requirement given in Section 2.2, because similar characteristic methods can often be formulated for different sets of governing equations. The equations of ideal magnetohydrodynamics (see, e.g., Kulikovsky and Lyubimov, 1965) and ideal compressible flow (Moretti and Abbott, 1966), for example, can be formulated and solved by characteristics. These solutions are often the best that can be obtained because the discontinuities which arise naturally in these physical systems are followed individually like the particles in the previous methods, and the continuous fluid behavior between the characteristics of the discontinuities is easier to represent. Even though conservation and positivity are relatively easy to ensure, characteristic methods are not generally applicable. The presence of even one diffusion term in the governing equations usually abrogates the use of characteristic methods because the characteristics of the model cease to exist.

A further drawback to characteristic methods is their complexity and relative inefficiency for complicated flows. Special precautions have to be taken to correctly describe shock overtaking, shock formation by the steepening of a compression wave, and the formation of a contact discontinuity when two shocks collide—in other words, all situations where discontinuities are introduced or removed. As Hoskin (1964) says, "the logical complications of the characteristic method increase rapidly with the number of discontinuities present, whereas finite-difference methods ... can deal with complicated systems as easily as with simple problems ... The advantages (of clarity and physical reality) of a characteristic grid get swamped by the difficulties of keeping account explicitly of all the special cases which may arise."

In two spatial dimensions the characteristics are surfaces which intersect along curved lines; in three dimensions they are solid figures (in four-space) intersecting along surfaces. The technique is conceptually much more complicated, and more complicated to program as well. The characteristics can become progressively more and more difficult to follow as the flows proceed in time. The relative inefficiency, which is not troublesome in simple flows where a few characteristics may be sufficient to describe the whole system, becomes far more serious when the new parallel and pipeline computers are contemplated (see Chapter 8). Since the characteristics, like quasiparticles, must each be considered individually, and since aspects of random access to computer memory as well as complicated logic are involved, the future of these methods on the newer computers is correspondingly dim.

Finally, characteristic methods are applicable to compressible systems rather than to incompressible systems where the sound speed is effectively infinite. In incompressible fluid flow problems, characteristic techniques reduce to the quasiparticle methods described above or to very similar algorithms.

2.5 Finite-Difference Methods

Unlike the quasiparticle and characteristic methods, finite-difference solutions of the convective equation are based on approximations in which the continuous flow variables such as $\rho(x,t)$ and $v(x,t)$ are discretized. A finite set of representative values $\{\rho_i\}$, $\{v_i\}$ is defined at N distinct points in space called grid (or mesh) points. The desired continuous functions are assumed to be known only at the mesh points, and a whole arsenal of techniques has been developed to predict future values of $\{\rho_i\}$, at some time t, given $\{\rho_i\}$ at $t = 0$. The term "finite differences" refers to the fact that the spatial derivatives (and usually the time derivative as well) are approximated by using only the grid point values, e.g.,

$$\left.\frac{\partial \rho}{\partial x}\right|_{x_i} \approx \frac{\rho_{i+1} - \rho_{i-1}}{x_{i+1} - x_{i-1}}. \tag{2.35}$$

The attraction of these finite-difference methods is their richness and simplicity. The classic text is by Richtmyer and Morton (1967), but the literature is vast and its growth shows no sign of stopping.

Although a detailed classification of possible finite-difference schemes is probably purposeless and may be impossible (there are too many hybrids), a few of the design choices should be discussed. The first of these is the distinction between Lagrangian and Eulerian methods. A Lagrangian finite-difference scheme (see, e.g., Steinberg, Kidder, and Cecil, 1966; Brennan and Whitney, 1970; Crowley, 1971; Chan, 1974) is one in which the location of the grid points is allowed to move along with the fluid, i.e., along the C_0 characteristic (2.34). This approach is useful because the convective term is transformed away in the moving system. Hence the positivity property is easily ensured, along with conservation. The trouble with this approach is that complicated flows in two or three dimensions will rather quickly distort the moving mesh.

Much of the accuracy is lost, often before useful information can be extracted from the calculation. There is another difficulty when Lagrangian methods are applied to the primitive equations. This stems from the fact that, in (2.13), the effective velocity with which energy is transported is $\mathbf{v}^* = \mathbf{v}(p + \mathscr{E})/\mathscr{E}$. Near stagnation points $\mathbf{v}^*$ can be up to twice as large as $\mathbf{v}$. Although the mesh comoves with the fluid mass, energy (and momentum) flow past just as surely as if the mesh were fixed.

In their original formulation Eulerian methods keep the grid fixed in position, so that distortion of the mesh cannot occur. Since fluid is now flowing across a stationary and often rather coarse mesh, truncation errors occur. These are secular in time and can play havoc with accuracy and positivity in a long run (Richtmyer and Morton, 1967; Wurtele, 1961; Roache, 1972; Boris and Book, 1973). Such errors are most serious and most likely to arise for compressible flows in the neighborhood of shocks, contact discontinuities, and sharp gradients. Thus they can be minimized by the use of a nonuniform grid which concentrates the mesh points in these critical regions. Since the latter move in the course of the calculation, the finely gridded portion of the mesh must migrate with them. For a rectilinear mesh in multidimensions, however, this cannot be done in general without deforming the mesh. Consequently, rezoning an Eulerian mesh is a compromise which can improve resolution in some places where it is needed, but runs the risk of overgridding in some places and undergridding in others. When it is possible to align the mesh with one feature of chief interest (e.g., a flame front) and regrid only with respect to the coordinate normal to this, the technique is very successful (Oran, Young, and Boris, 1980). The other way around the restriction on completely adaptive regridding is to abandon rectilinearity, for example, by using triangles instead of rectangles. Of course, the adaptive gridding approach, carried to its limit, can give rise to a fully Lagrangian scheme, but that is not necessarily the best way to develop such a scheme. Chapter 6 presents an example of a fully Lagrangian two-dimensional code based on triangular gridding which uses reconnections instead of rezoning.

Boris and Book (1976) have shown that in Eulerian finite-difference schemes there is an irreducible residual error associated with the Gibbs effect. To see this, consider an approximation to (2.1) (with $v = $ const) having *no* phase and amplitude errors. This can be realized on a mesh with uniform spacing δx by Fourier–analyzing the initial profile $\rho_0(x) = \rho(x,0)$, multiplying each harmonic by a phase factor $\exp(-ikvt)$, where k is the wavenumber, and then Fourier–synthesizing. At times $t_n = n\delta x/v$, where n is an integer, the solution coincides with the result

$$\rho_j^n = \rho(x_j,t_n) = \rho_0(x_j - vt_n). \tag{2.36}$$

At intermediate times, however, "wiggles" appear due to the discreteness of the Fourier transform (Fig. 2-1). As is clear, they can even destroy positivity. They are purely a consequence of the finiteness of the discrete representation—

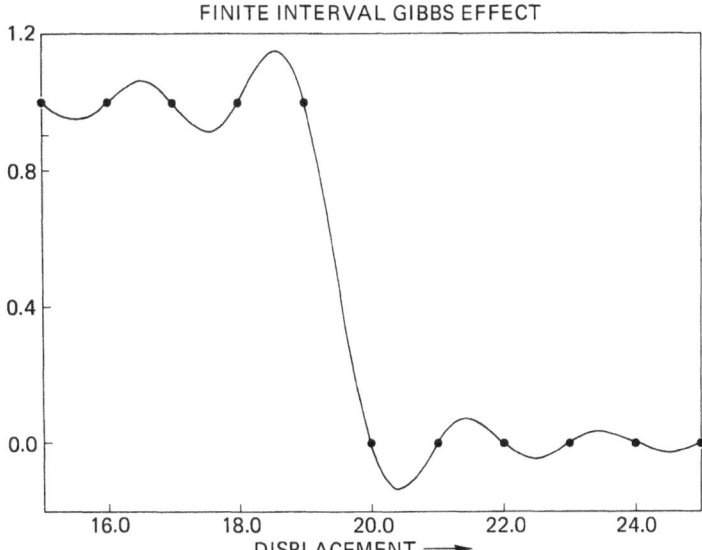

Fig. 2-1 The Gibbs phenomenon at the edge of a square wave. The points show the numerical values of density for the leading edge of a square wave with transition from 0 to 1. The solid line is the continuous finite Fourier expansion which passes through all the grid values. This Fourier expression, the smoothest function passing through all the grid values, displays overshoots and undershoots. When it is advected over a nonintegral multiple of the mesh space, the undershoots can cause the density to become negative on some grid points.

a manifestation of the "uncertainty principle" mentioned in Section 2.2. No recourse to higher-order accuracy, even the "infinite-order" accuracy of spectral methods, can eliminate them; rather, the cure is to adopt a nonlinear algorithm with the positivity or "monotonicity" property built in, such as those described in Chapter 3.

The second major finite-difference choice is that of the time integration technique, in particular the controversial question of choosing an explicit or an implicit scheme, or some mixture of the two. Implicit schemes for the convective derivatives involve, in the worst cases, gigantic matrix equations, and in the best cases, the inversion of a tridiagonal matrix. Thus their operation count and speed are appreciably worse than that obtained using the corresponding explicit schemes. The advantage of implicit schemes, which in some problems outweighs the dual disadvantages of program complexity and operation count per time step, is the ability to take much longer finite-difference time steps without exciting numerical modes of instability. Chapter 4 provides an illustration of how advantageous it can be to use a combination of explicit and implicit schemes.

The choice between explicit and implicit differencing lies in the physics (or the mathematics) of the specific problem being solved. When the physical phenomenon of interest varies appreciably on a time scale τ, no calculation with $\Delta t \gg \tau$ can reasonably be claimed to reproduce the phenomenon accurately. In such cases computationally efficient explicit methods are called for, with $\Delta t < \tau$ for stability. For example, suppose $\mathbf{v}$ is constant in (2.1). In one dimension the simple upwind (donor-cell) finite-difference approximation to (2.1) for $v > 0$ on a uniform mesh is

$$\frac{\rho_i^{n+1} - \rho_i^n}{\delta t} = -v \frac{\rho_i^n - \rho_{i-1}^n}{\delta x} \tag{2.37}$$

where n labels the time level. A sinusoidal profile $\rho_j^n = e^{ijk\delta x}$ experiences an amplification $\mathscr{A}$ in a single cycle, where

$$\mathscr{A}^2 = (1 - \varepsilon)^2 + \varepsilon^2 + 2\varepsilon(1 - \varepsilon)\cos k\delta x; \tag{2.38}$$

here we have written $\varepsilon = v\delta t/\delta x$. The corresponding implicit differencing formula,

$$\frac{\rho_i^{n+1} - \rho_i^n}{\delta t} = -v \frac{\rho_i^{n+1} - \rho_{i-1}^{n+1}}{\delta x}, \tag{2.39}$$

yields an amplification $\mathscr{A}'$ given by

$$\mathscr{A}'^2 = [(1 + \varepsilon)^2 + \varepsilon^2 - 2\varepsilon(1 + \varepsilon)\cos k\delta x]^{-1}. \tag{2.40}$$

$\mathscr{A}$ and $\mathscr{A}'$ differ by terms of order ε^2, and both differ from unity, the correct analytic result, by terms of order ε. When $\varepsilon > 1$, $\mathscr{A} > 1$ for sufficiently large $k\delta x$, implying that the explicit numerical approximation is unstable; in contrast, $\mathscr{A}' \leq 1$ for all values of ε and $k\delta x$. But note that for $\varepsilon > 1$ the error in the implicit scheme, which is $O(\varepsilon)$, is at least of order unity. In other words, the stability condition for an explicit method is usually the accuracy condition for the corresponding implicit method.

When the phenomenon of real interest is not the fastest (shortest time scale) effect in the problem, additional considerations are operative. Consider, for example, the expanding flow of compressed air from a whistle. The flow speed is slow, so the fine-scale high-speed vibrations of the whistle sound are expected to have little effect on the slow (time-averaged) macroscopic flow pattern, even though the energy emitted as sound may be comparable with the flow energy imparted to the air. Clearly an implicit scheme is desired since the speed of sound is about two orders of magnitude larger than the flow speed, and the condition $\Delta t \lesssim \tau_{\text{flow}}$ is far less stringent than the more expensive $\Delta t \lesssim \tau_{\text{sound}}$ found for explicit schemes. Unfortunately the problem is not yet resolved because implicit schemes which are forward differenced beyond second-order centered differencing tend to damp the undesired high-frequency sound waves strongly. Thus sonic energy from the whistle would be deposited as heat near the whistle. A centered scheme (or one quite near centered) would allow much of the sonic energy to propagate away as sound without heating the local air. The short-wavelength sonic components are strongly dispersed, however. They would generally stay in the vicinity of the whistle far too long and might eventually deposit their energy near the whistle anyway.

We will see that the problems with both the Eulerian and Lagrangian approaches can be largely overcome. Finite-difference methods have solved the transient ideal Rankine–Hugoniot shock problem adequately. An example of a general purpose finite-difference algorithm for solving compressible flow problems is ETBFCT, discussed in Section 3.2. Characteristic methods are also adequate, as are some quasiparticle methods when enough particles are used. None of the other methods has been developed sufficiently to make such calculations routine. Until these become commonplace and generally attractive, finite-difference methods would seem to have the inside track for general applications to compressible flow problems.

For incompressible hydrodynamics problems formulated in terms of the primitive equations the best solution involves developing the Poisson equation (2.10) for the fluid pressure. This in effect implicitly couples all of the grid points to obtain a formulation without sonic or compression waves. Hence the pressure at any location is generally that value required to ensure that $\frac{d(\nabla \cdot \mathbf{v})}{dt} \equiv 0$. Of course cavitation leads to large changes in density because the pressure cannot go negative. In actual physical systems the $\nabla \cdot \mathbf{v} = 0$ condition relaxes before the $p \geq 0$ condition does. Within the Eulerian framework there are a number of ways to develop an implicit formulation. One of these is exemplified by the semi-implicit technique described in Chapter 4. Explicit formulations leave the treatment of the convective terms free for nonlinear monotonic integration (Boris and Book, 1973, 1976; Harten, 1974; Van Leer, 1977, 1977a; Hain, 1978), which is much more accurate than the linear nonpositive algorithms usually used. Section 3.3 describes a nonlinear monotonic Eulerian finite-difference scheme adapted for use in connection with incompressible flows.

Finite-difference methods have been available since the early 1960's which clearly satisfy our requirements (1), (2), and (6) (Roache, 1972). Nonlinear algorithms like those described in Chapter 3 in addition meet requirement (3). Requirements (4) and (5), however, confront the user with the necessity to define his objectives: how accurate? and how efficient? Certainly for some problems, spectral, characteristic, or other techniques are demonstrably more accurate than finite differences. It is our experience, however, that this is usually achieved at the price of a commensurate increase in running times or loss of flexibility. On the other hand, trying to save computing costs by reducing accuracy in a finite-difference treatment also can be counterproductive. For almost all problems of interest, results of acceptable accuracy require a finite-difference algorithm which is second-order accurate in both space and time. Kreiss (1972) goes farther and concludes that if Eulerian difference methods are used, they should be at least fourth-order accurate (in space).

2.6 Finite-Element Methods

Both finite-element (Strang and Fix, 1973; Gottlieb, 1972) and spectral
(Gottlieb and Orszag, 1977) methods utilize continuum concepts fully in
solving the fluid equations. Both methods are based on transforms. The fluid
variables are expanded in a set of useful basis functions whose expansion
coefficients are determined from a complicated set of coupled, usually
nonlinear, but ordinary differential equations. For finite elements the basis
functions are polynomials localized to a mesh cell or two, and the techniques of
matching expansion function boundary conditions at cell boundaries resemble
those involved in using splines, whereas spectral methods generally employ
global basis functions. Both methods employ projection to recover ordinary
differential equations which are implicit in one way or another. The finite-
element methods acquire their· implicitness by coupling spatially adjacent
basis functions via projection when evaluating, for example, the time derivative
on the left-hand side of (2.9).

Gottlieb and Orszag (1977) distinguish three different kinds of transform
methods: the Galerkin, tau, and collocation (pseudospectral) approximations.
To understand how they are defined, consider an initial value problem of the
form

$$\frac{\partial u(x,t)}{\partial t} = \mathscr{L}(x,t)[u(x,t)] + f(x,t), \tag{2.41}$$

where the operator $\mathscr{L}$ is in general nonlinear. We assume that boundary
conditions of the form

$$B_{\pm}(x)u(x,t) = 0 \tag{2.42}$$

hold at $x = x_{\pm}$, where the $B_{\pm}$ are linear operators and

$$u(x,0) = u_0(x). \tag{2.43}$$

We approximate u by a superposition of N basis functions $v_n(x)$ satisfying the
boundary condition (2.42),

$$u(x,t) = \sum a_n(t)v_n(x). \tag{2.44}$$

If $\{v_n(x)\}$ form an orthonormal set, the coefficiénts a_n satisfy

$$a_n(t) = \int_{x_-}^{x_+} dx v_n(x)u(x,t). \tag{2.45}$$

Equation (2.41) then yields

$$\frac{da_n}{dt} = \sum_{m=1}^{N} \int_{x_-}^{x_+} dx v_n(x)\mathscr{L}v_m(x) + \phi_n(t), \tag{2.46}$$

where ϕ_n is defined by

$$f(x,t) = \sum \phi_n(t)v_n(x). \tag{2.47}$$

Equation (2.46), an explicit equation for da_n/dt, is especially simple by virtue of our assumption that $\{v_n(x)\}$ are orthonormal. In the general case, it is necessary to solve simultaneously the N linear equations

$$\sum_{m=1}^{N} \left\{ \frac{da_m}{dt} \int_{x_-}^{x_+} dx v_n(x) v_m(x) - \int_{x_-}^{x_+} dx v_n(x) \mathcal{L} v_m(x) \right\} = \phi_n. \qquad (2.48)$$

Together (2.44) and (2.48) constitute the Galerkin approximation.

The tau approximation (Lanczos, 1956) starts again with (2.44), but the functions $v_n(x)$ are not required to satisfy the foundary conditions. Instead, the two constraints,

$$\sum_{n=1}^{N} a_n B_\pm v_n(x) = 0, \qquad (2.49)$$

are applied. To leave the problem well posed, only the first $N - 2$ of Eqs. (2.48) [or (2.46) if the basis functions are orthonormal] are retained.

For collocation methods, N points $\{x_N\}$ are chosen such that

$$x_- < x_1 < \cdots < x_N < x_+. \qquad (2.50)$$

The solution $u(x)$ of (2.41) is approximated by (2.44) where the coefficients a_n satisfy

$$\sum_{m=1}^{N} a_m v_m(x_n) = u(x_n), \qquad (2.51)$$

$n = 1, 2, \ldots, N$. For example, if $\{x_n\}$ are uniformly distributed between x_- and x_+ and $\{v_n\}$ are harmonic functions (sines and cosines), (2.51) just defines a discrete Fourier transform.

Note that all three of these approximations are formal, in the sense that we have not yet specified the $\{v_n\}$ nor supplied a prescription for carrying out the time integration. In each case, though, the formulation converts the partial differential (or integro-differential) equation (2.41) we began with into a set of ordinary equations, for which standard integration packages exist.

In developing the finite-element technique one can work with any of the three approximation techniques. It is convenient to use the Galerkin approximation (2.48) as the starting point. The system becomes linear if $\mathcal{L}$ is linear, in which case an explicit or implicit finite-difference approximation to the time derivative allows the $\{a_n\}$ to be updated by matrix algebra techniques. Linearity is obviously not necessary in general, however.

Equations (2.48) specify that the error

$$\varepsilon(v, t) = \frac{\partial u}{\partial t} - \mathcal{L}[u] - f \qquad (2.52)$$

be orthogonal to each of the basic functions v_n, which is equivalent to the requirement that the integral of ε^2 be minimized on the interval $x_- < x < x_+$. This provides the foundation for developing a systematic way of estimating the accuracy of the approximation to (2.41).

In the finite-element technique, each of the functions v_n vanishes outside a domain D_n, and the $\{D_n\}$ comprise a covering of the domain of definition $D = (x_-, x_+)$. Hence if the $\{D_n\}$ are nonoverlapping, the $\{v_n\}$ are automatically orthogonal to one another. The $\{v_n\}$ are chosen in some simple form to facilitate evaluation of the integrals. For example, the "tent" or "roof" functions may be used:

$$
\begin{aligned}
v_n(x) &= \frac{x - x_{n-1}}{x_n - x_{n-1}}, & x_{n-1} \le x \le x_n; \\
&= \frac{x_{n+1} - x}{x_{n+1} - x_n}, & x_n \le x \le x_{n+1}; \\
&= 0 & \text{otherwise,}
\end{aligned}
\tag{2.53}
$$

where the $\{x_n\}$ satisfy (2.50). The corresponding domains are evidently $D_n = (x_{n-1}, x_{n+1})$.

It is straightforward to generalize this method to multidimensional problems in several dependent variables. The domain D_n need not have regular shapes, so that the method is well suited to problems with irregular boundaries. The domain can also change in the course of time to allow improvements in resolution where they are needed. An interesting one-dimensional adaptive version of the finite-element technique has been developed by Gelinas, Doss, and Miller (1979). They use the tent functions (2.53), allowing the node positions $\{x_n\}$ to move in time, so that the time derivative of $u(x,t)$ is given by

$$
\frac{\partial u}{\partial t} = \sum \left[\dot{a}_n(t) v_n(x) + \dot{x}_n(t) w_n(x) \right],
\tag{2.54}
$$

where

$$
\begin{aligned}
w_n(x) &= -\frac{a_n - a_{n-1}}{x_n - x_{n-1}} \frac{x - x_{n-1}}{x_n - x_{n-1}}, & x_{n-1} \le x < x_n; \\
&= -\frac{a_{n+1} - a_n}{x_{n+1} - x_n} \frac{x_{n+1} - x}{x_{n+1} - x_n}, & x_n < x \le x_{n+1}; \\
&= 0 & \text{otherwise.}
\end{aligned}
\tag{2.55}
$$

It follows that all integrals over D of the products $v_n v_m$, $v_n w_m$, and $w_n w_m$ vanish except when $n = m$ or $n - m = \pm 1$. Instead of (2.48) a set of $2N$ ordinary differential equations in terms of $\dot{a}_1, \ldots, \dot{a}_N$ and $\dot{x}_1, \ldots, \dot{x}_N$ results. These can be expressed in the form

$$
\begin{aligned}
\sum_m \left[(v_m, v_n)\dot{a}_m + (v_m, w_n)\dot{x}_m \right] &= (\mathscr{L}[u] + f, v_n), \\
\sum_m \left[(w_m, v_n)\dot{a}_m + (w_m, w_n)\dot{x}_m \right] &= (\mathscr{L}[u] + f, w_n),
\end{aligned}
\tag{2.56}
$$

$n = 1, \ldots, N$, where we have defined the inner product of two functions g_1, g_2 by

$$(g_1, g_2) = \int_{x_-}^{x_+} dx g_1(x) g_2(x). \tag{2.57}$$

The conventional fixed-node Galerkin equations (2.48) are recovered from (2.56) by setting $w_n = 0, n = 1, \ldots, N$. In matrix form (2.56) can be written

$$\mathfrak{M} \cdot \dot{\mathbf{b}} = \mathbf{s}, \tag{2.58}$$

where $b_n = a_n, b_{N+n} = x_n, n = 1, \ldots, N$.

At any point x_n such that

$$\frac{a_n - a_{n-1}}{x_n - x_{n-1}} = \frac{a_{n+1} - a_n}{x_{n+1} - x_n} \tag{2.59}$$

the function w_n equals a multiple of v_n, and the matrix M becomes degenerate. This reflects the fact that the profile of the approximation (2.44) to u becomes a straight line through the three points (x_{n-1}, a_{n-1}), (x_n, a_n), and (x_{n+1}, a_{n+1}), on which the middle point can be relocated arbitrarily. The singularity can be removed by minimizing, instead of $(\varepsilon, \varepsilon)$, the quantity $(\varepsilon, \varepsilon) + \gamma \sum_{n=1}^{N-1} (\dot{x}_{n+1} - \dot{x}_n)^2$. The effect of doing this is to relocate nodes in a way that penalizes relative motion. This algorithm is central to the "Moving Finite Element" technique and is responsible for producing adaptive "regridding" in an efficient and general way. Nodes migrate toward zones where abrupt changes occur in the dependent variable. Even though $2N$ variables are used instead of N, this reduces the total machine time in many cases. The method has not yet been generalized to multidimensions, but has been successful in applications to one-dimensional combustion problems.

Finite-element methods can equally well be based on tau or collocation approximations. Thus far they have not been exploited for fluid calculations as fully as have quasiparticle, finite-difference, and spectral methods, so a final verdict on them cannot be reached. In general the number of operations per node exceeds that per mesh point in finite-difference methods, so that finite-element methods will have to prove significantly more accurate than the latter if they are to gain increased acceptance.

2.7 Spectral Methods

Spectral methods for solving initial-value partial differential equations involve representing the dependent variables as linear superpositions of a set of basis functions of the independent variables. These functions are chosen so that their properties (e.g., analyticity, orthogonality, the property of satisfying boundary conditions, that of possessing some close connection with the physical processes in the problem, etc.) simplify the task of decomposition, i.e., calculating the series coefficients. The best known and most useful such representation is in terms of sinusoidal functions, i.e., Fourier analysis.

The process of superposition and its inverse, decomposition, involve all N basis functions. In contrast, second-order difference approximations involve just three points at a time. Consequently, to be competitive, spectral methods require very efficient transform methods. Such a method in the case of Fourier transforms is the fast Fourier transform (FFT), developed first by Cooley and Tukey (1965). Fast transform methods have been found for other orthogonal bases (Orszag, 1979), as will be seen. Like finite-element techniques, spectral techniques can be based on any of the three general transform approximation procedures described in the previous section.

The time derivative may be differenced either explicitly or implicitly. The most popular explicit scheme is simple leapfrog. Runge–Kutta methods are also used. If these are of third or higher order, the linear stability condition has the same form as that for leapfrog,

$$\frac{(N - 1)\bar{v}\Delta t}{x_+ - x_-} < C, \tag{2.60}$$

where $\bar{v}$ has the significance of an average or effective velocity and C is a constant of order unity which depends on the scheme. (Note the similarity to the Courant condition in finite differences.)

Among implicit schemes, the Crank–Nicholson and backward Euler schemes are stable for hyperbolic equations without restriction on the time step. Of course accuracy is another story. Even unconditionally stable schemes must satisfy (2.60) to be accurate.

All three spectral techniques can be generalized to multidimensions without difficulty. Clearly a large selection of different spectral approximations is possible, depending on which kind of technique is adopted, the choice of basic functions, and the order N of the approximation. These should reflect the physics of the problem and the accuracy requirements imposed on the desired solution. As always, an important consideration to be balanced against the latter is running time. This is mainly determined by the time required to perform the transform (2.45) and its inverse (2.44).

Orszag (1979) has concluded that the basis set should be chosen to consist of eigenfunctions of nonsingular Sturm–Liouville equations (e.g., sinusoidal functions) only for problems whose boundary conditions match those of the Sturm–Liouville problem. In all other cases, the basis should consist of eigenfunctions of a singular Sturm–Liouville problem, i.e., those for which the eigenvalue condition is that the solution remain well behaved at some boundary. (Chebychev and Legendre polynomials are the most familiar examples.) This prescription is made to ensure that the Gibbs phenomenon does not arise in the solution at the boundary.

Such bases give rise to eigenvalue problems

$$\mathfrak{L} \cdot \mathbf{u} = \mathbf{F} \tag{2.61}$$

whose spectral approximation

$$\mathfrak{L}_s \cdot \mathbf{u}_s = \mathbf{F}_s \tag{2.62}$$

involves a full $N \times N$ matrix $\mathfrak{L}_s$. In a three-dimensional ($\sim 100 \times 100 \times 100$) problem, $N \approx 10^6$. Since storage of the matrix requires N^2 words and straightforward direct inversion involves N^3 operations, obviously recourse must be had to another technique (Orszag, 1979). The trick is to find a sparse approximate operator $\tilde{\mathfrak{L}}$ such that $\|\tilde{\mathfrak{L}}^{-1} \cdot \mathfrak{L}_s\| = O(1)$ as $N \to \infty$. If $\tilde{\mathfrak{L}}$ can be inverted cheaply (see Section 7.1), then an iteration scheme, e.g.,

$$\tilde{\mathfrak{L}} \cdot \mathbf{u}^{\text{new}} = \tilde{\mathfrak{L}} \cdot \mathbf{u}^{\text{old}} - \omega(\mathfrak{L} \cdot \mathbf{u}^{\text{old}} - \mathbf{F}), \tag{2.63}$$

where ω is a relaxation constant, yields machine accuracy in a finite number of iterations. Here $\tilde{\mathfrak{L}}$ is chosen to be sparse so that storage requires only $O(N)$ words, and at the same time $\tilde{\mathfrak{L}}$ is efficiently invertible. In practice $\tilde{\mathfrak{L}}$ can be a finite-difference approximation, invertible in $O(N)$ operations. Aside from these, the total operation count is proportional to $N \ln N$. (It should be noted, however, that the constant of proportionality is much larger for, e.g., Bessel or Bernoulli functions than it is for the conventional FFT.)

Spectral methods are applicable to nonlinear problems, but the analysis of stability, accuracy, etc., of course becomes nontrivial. Following Gottlieb and Orszag (1977), we consider the problem of solving the incompressible Navier–Stokes equation. This is written in the vorticity–stream-function formulation. The difficulties arise from the nonlinear term in (2.46). Collocation is recommended for such problems. The key is to evaluate the derivative $\partial u / \partial x$ (where now u denotes the vorticity ζ) using the fast transform. For complex exponential basis functions,

$$\frac{\partial u}{\partial x} = \frac{2\pi}{x_+ - x_-} \sum in \, a_n \exp\left(\frac{2\pi in \Delta x}{x_+ - x_-}\right). \tag{2.64}$$

Then the solution is marched forward in configuration space at the collocation points. The number of operations for this procedure again goes as $N \ln N$. (We will see an example of this technique applied to another problem shortly.)

Problems involving irregular boundaries (i.e., anything other than squares and circles) can be solved by conformal mapping. There is nothing hard about this in principle, but each reformulation or dynamical motion that results in a change in boundary conditions or topology (or extreme changes in geometry) necessitates recoding "by hand."

Provided no extreme gradients occur, spectral codes require much less resolution than finite-difference schemes (since they have in effect infinite-order accuracy) and become proportionately more competitive. If the location of a steep gradient is known *a priori* (for example, in a boundary layer), they can concentrate the resolution there. Without the use of nonlinear techniques analogous to those employed in finite-difference schemes to guarantee positivity (Chapter 3), spectral techniques normally cannot resolve moving near-discontinuities, e.g., at shock fronts and breaking internal waves.

Recently, however, Gottlieb and Orszag (1979) have demonstrated good agreement with analytical solutions in several test problems involving shocks of moderate strength (i.e., with Mach number $M \sim 2$), using a collocation technique and Chebychev expansions. To eliminate the Gibbs-effect wiggles (described in Section 2.5), they applied a low-pass filter to the fluid variables every ~ 200 time steps, and an additional cosmetic filter ("postprocessor") prior to displaying the profiles. The shocks obtained using $N = 64$ terms in the Chebychev expansion were resolved to within $\sim 1/40$ of the total system size, equivalent to $\sim 1\frac{1}{2}$ mesh spaces in a finite-difference scheme. Remarkably, contact discontinuities were resolved equally well.

The theoretical explanation for this success with essentially linear filters is associated with the idea that the spurious oscillations contain information from which it is possible to deduce the correct (nonoscillatory) solution. The principal computational difficulties experienced in implementing the method were connected with circumventing the stringent time-step limitation ($\delta t < \delta t_{\max} \sim 1/N^2$) caused by the high resolution yielded by Chebychev expansions near the boundaries (which was not even needed in the test problems) and with determining consistent boundary conditions.

In many problems of interest it is possible by means of asymptotic analysis to derive an approximate nonlinear equation involving a single fluid quantity, e.g., one component of the flow field. Typically it is necessary to assume for this purpose that the amplitude of the disturbance, though finite, is small and the horizontal scale is large compared with the vertical scale. Such a nonlinear wave equation is useful because, in addition to decreasing the number of dependent variables, it reduces the number of spatial dimensions by requiring the vertical dependence to be that of a linear eigenmode. Thus, a three-dimensional process may sometimes be approximated as a two-dimensional one; more commonly, a two-dimensional process is approximated by a one-dimensional wave equation.

Examples of such equations are the Korteweg–deVries (KdV) equation (Scott, Chu, and McLaughlin, 1973) for surface water waves and the Benjamin–Ono equation (Benjamin, 1967) for internal waves. Many of these equations are amenable to exact solution by the inverse-scattering method (Scott, Chu, and McLaughlin, 1973; Ablowitz et al., 1974). Among the solutions so obtained, the most remarkable are those describing solitons. These are isolated bulges (or troughs), rather than trains of waves, and have the property of retaining their identity even when two or more collide and pass through one another. (Clearly this property is irrelevant to the characterization of a single isolated soliton.)

It is important to note, however, that solitons may result even if the inverse scattering technique and other analytic techniques are inapplicable; indeed, solitons were first observed by Zabusky and Kruskal (1965) in numerical solutions of the KdV equation. More significantly, solitons are seen in nature and in solutions of "exact" hyperbolic systems, i.e., systems of equations like (2.1)–(2.3) which have not been treated with the weak-nonlinearity long-wavelength approximations. Thus solitons are more general and more fundamental than our analytic techniques for deriving them. This means that for many purposes, simulation is an indispensable tool in their investigation.

For integrating nonlinear wave equations, a variety of techniques has been employed. Although explicit finite-difference schemes were used in the early numerical work (e.g., Zabusky, 1967) and to solve a modified form of the KdV equation having a dissipative term and a linear forcing term (Ott *et al.*, 1973), the most efficient and accurate methods are spectral or quasispectral.

A straightforward technique for finding periodic solutions is to work in k space. For example, the Benjamin–Ono equation becomes

$$a\frac{\partial v_k}{\partial t} + b\sum_{k'} ik'v_{k'}v_{k-k'} + ik|k|cv_k = 0, \tag{2.65}$$

where

$$v(x,t) = \sum_k e^{ikx}v_k. \tag{2.66}$$

Equation (2.65) can be treated as an array of ordinary differential equations, with a linear or quadratic driving term added (Ott *et al.*, 1973). Even though explicit, this technique is efficient when FFT's are used. Moreover, it is highly accurate because the spatial derivatives are Nth-order accurate, where N is the number of modes retained in the sum (2.66). It has the additional advantage of being easily convertible into a solution of the two-dimensional problem, provided one is willing to accept a limitation on the number of modes so that the matrix multiplication operation remains within manageable size.

Perhaps the best technique developed to date is the split-step Fourier method of Tappert (1974). [See also the discussion by Hyman in Lax (1975).] Suppose u satisfies

$$\frac{\partial u}{\partial t} + \frac{\partial}{\partial x}f(u) + \mathscr{L}\frac{\partial u}{\partial x} = 0, \tag{2.67}$$

where f is some function of u and $\mathscr{L}$ is a linear differential operator with constant coefficients,

$$\mathscr{L} = \sum a_n\frac{\partial^n}{\partial x^n}. \tag{2.68}$$

Let $\mathscr{F}$ denote the Fourier transform operation. Then

$$\mathscr{F}(\mathscr{L}u) = L(k)\mathscr{F}(u), \tag{2.69}$$

where

$$L(k) = \sum a_n (ik)^n. \tag{2.70}$$

An important feature of the method, of course, is the use of fast transform routines for $\mathscr{F}$.

To go from time t to $t + \delta t$, we proceed in two steps. First, advance the solution using only the nonlinear term. An implicit finite-difference scheme is used for this purpose:

$$\tilde{u}(t + \delta t) = u(t) - \tfrac{1}{2}\delta t\{\mathscr{D}[\tilde{u}(t + \delta t)] + \mathscr{D}[\tilde{u}(t)]\} \tag{2.71}$$

where $\mathscr{D}(u)$ approximates $\partial f(u)/\partial x$. Then advance the solution exactly, using only the linear term, according to

$$u(t + \delta t) = \mathscr{F}^{-1}[e^{ikL(k)\delta t}\mathscr{F}(\tilde{u})]. \tag{2.72}$$

This method is second-order accurate and linearly stable.

One concludes that one-dimensional nonlinear wave equations require careful treatment of the (nonlinear) advective term, but do not suffer much from errors in approximating the time derivative. In this connection, Fornberg (1975) has shown that even high-order finite-difference approximations to the advection term are qualitatively less accurate than spectral methods, a result that seems specific to convective wave equations.

CHAPTER 3

Flux-Corrected Transport

There is undoubtedly some merit in trying to improve the performance of quasiparticle methods, on the one hand, and of transform methods, on the other, so as to attain the level of generality and utility enjoyed by finite differences without sacrificing accuracy. But patching up the obvious failings of the front-runners, finite-difference methods, is a low-risk high-return investment. In Lagrangian methods, the outstanding problems arise from secular distortions of the grid which quickly disrupt calculations of most interesting flows. In Eulerian methods, the major weakness in a large class of problems of interest is the need for a large artificial damping (numerical diffusion) to fill in what would otherwise be pits of "negative density" in the calculated profiles. Since the "Eulerian" positivity problem is encountered even in Lagrangian calculations for many situations, it demands attention.

3.1 Improvements in Eulerian Finite-Difference Algorithms

The outstanding problem with Eulerian techniques for solving continuity equations is the competition between accuracy and positivity (Kreiss, 1972; Van Leer, 1974). Consider the rather general three-point approximation to (2.1):

$$\tilde{\rho}_i = \rho_i^o - \tfrac{1}{2}(\rho_{i+1}^o + \rho_i^o)\varepsilon_{i+1/2} + \tfrac{1}{2}(\rho_i^o + \rho_{i-1}^o)\varepsilon_{i-1/2}$$
$$+ v_{i+1/2}(\rho_{i+1}^o - \rho_i^o) - v_{i-1/2}(\rho_i^o - \rho_{i-1}^o) \tag{3.1}$$
$$= \rho_i^o - \Delta x_i^{-1}[F_{i-1/2} - F_{i+1/2}],$$

where $\varepsilon_{i+1/2} = v_{i+1/2}\delta t/\delta x_{i+1/2}$. Equation (3.1) is in finite-difference conservation form, with superscript o used to distinguish "old" values of ρ, whole subscript indices representing cell centers, and half indices indicating cell interfaces. At every cell i, the derived quantity $\tilde{\rho}_i$ differs from ρ_i as a result of the inflow and outflow of quantities of material $F_{i\pm1/2}$. These quantities, called fluxes, are successively added and subtracted all along the array of densities ρ_i, so that the conservation condition is satisfied explicitly (subject to behavior at the boundaries). The expressions involving $\varepsilon_{i\pm1/2}$ are called transportive fluxes; these terms by themselves yield a first-order approximation to Eq. (2.18). The additional numerical diffusion terms with diffusion coefficients $v_{i+1/2}$ have to be added to ensure positivity. The stability of (3.1) is ensured, at least roughly, when

$$\tfrac{1}{2} > v_{i+1/2} > \tfrac{1}{2}\varepsilon_{i+1/2}^2. \tag{3.2}$$

The upper limit arises from the explicit diffusion time-step condition, while the lower limit is the Lax–Wendroff damping coefficient (Boris and Book, 1976a). This lower limit can be written in the equivalent form

$$v_{i+1/2} = \tfrac{1}{2}|\varepsilon_{i+1/2}|(|\varepsilon_{i+1/2}| + c), \tag{3.3}$$

where we have introduced a positive "clipping factor" c. When $c = 0$, scheme (3.1) is of second order. Unfortunately, positivity is only ensured linearly when $c \sim 1$, whereupon (3.1) reduces to the donor-cell scheme.

The escape route is signaled in the preceding sentence by the word "linearly." By relaxing the linearity implied by (3.1) and letting the diffusion coefficients be nonlinear functionals of the quantities ρ and ε, we can hope to reduce the integrated dissipation below the rather severe limit $c = 1$, and yet retain sufficient dissipation near steep gradients to ensure positivity. Boris and Book (1973) introduced this basic nonlinear approach with the techniques of flux-corrected transport (FCT). A literature is beginning to form (Boris and Book, 1976, 1976a; Van Leer, 1974, 1976, 1976a; Harten, 1974; Book, Boris and Hain, 1975; Hain, 1978) about these "monotonic difference schemes" since the dilemma of accuracy versus positivity in Eulerian difference schemes can be resolved in no other way. Several different approaches are possible, and the area is still largely unexplored.

In its essentials the FCT method in one dimension consists of six sequential operations at each interior point x_i:

(1) Compute $F_{i+1/2}^L$, the transportive flux given by some low-order scheme guaranteed to give monotonic (ripple-free) results for the problem at hand.

(2) Compute $F_{i+1/2}^H$, the transportive flux given by some high-order scheme.

(3) Define the "antidiffusive flux":

$$A_{i+1/2} = F_{i+1/2}^H - F_{i+1/2}^L.$$

(4) Compute the updated low-order ("transported and diffused") solution:

$$\rho_i^{td} = \rho_i^o - \Delta x_i^{-1}[F_{i+1/2}^L - F_{i-1/2}^L].$$

(5) Limit the $\{A_{i+1/2}\}$ in a manner such that ρ^n as computed in step 6 is free of overshoots and undershoots:

$$A_{i+1/2}^C = C_{i+1/2}A_{i+1/2}, \qquad 0 \le C_{i+1/2} \le 1.$$

(6) Apply the limited antidiffusive fluxes to get the new densities ρ_i^n:

$$\rho_i^n = \rho_i^{td} - \Delta x_i^{-1}[A_{i+1/2}^C - A_{i-1/2}^C].$$

The critical step in the above is, of course, step 5, which will be discussed in detail in Section 3.3. The $\{A_{i+1/2}\}$ are chosen to ensure that the algorithm reduces, as nearly as is consistent with avoiding the introduction of dispersive ripples, to the higher-order scheme ($c = 0$). In the complete absence of the flux limiting step ($A^C_{i+1/2} = A_{i+1/2}$), ρ^n_i would simply be the time-advanced high-order solution. We note that this definition of FCT is considerably more general than that given originally by Boris and Book (1973).

In any Eulerian calculation, numerical diffusion arises because material which has just entered a computational cell, and is still near one boundary, becomes smeared over the whole cell. FCT minimizes this effect. If the fluid equations are written in conservative form, both schemes are implemented using transportive fluxes. The procedure for assigning weights involves limiting or "correcting" the fluxes at certain points. The weights for the low-order scheme are chosen so that near sharp discontinuities enough diffusion is supplied to automatically retain monotonicity. At shock fronts this amounts to turning on a local dissipation process.

A numerical diffusion Reynolds number $(Re)_{ND} = 2L/c\delta x$ can be defined, where L is the characteristic size of a structure in the flow. Even the most accurate (spectral) simulations require setting $c = 1$ to guarantee positivity linearly. This gives rise to the usual definition of the numerical Reynolds number, $2L/\delta x$. Algorithms such as FCT which guarantee monotonicity non-linearly can have average values $\langle c \rangle \sim 10^{-1}$–$10^{-2}$, introducing much less overall dissipation and permitting calculations with effective Reynolds numbers such that $Re \sim (Re)_{ND} \gg 2L/\delta x$.

Figure 3-1 shows a comparison of four common difference schemes solving the standard square wave advection problem. The effects of excess numerical damping in the donor-cell treatment (upstream centered first order) and of excess dispersion in the leapfrog and Lax–Wendroff treatments are clearly visible. Dispersion manifests itself as a trail or projection of oscillations in the computed solution near discontinuities and sharp gradients of the "correct" solution. The two second-order algorithms yield results which are almost indistinguishable. In Fig. 3-1a we have a result typical of those obtained from a first-order algorithm, even in more general contexts, when sharp gradients arise. Conversely, the results of Figs. 3-1b and 3-1c typify the dispersive errors to be expected in second-order treatments. The results obtained by using SHASTA, the fourth algorithm tested, show a striking improvement. The "discontinuities" are resolved to within a mesh space or two. At the same time, gentle gradients ("continuous" portions of the profile) are propagated with second-order accuracy. Thus the FCT algorithm is able to handle shocks and steep concentration gradients while incurring a minimal penalty in regions where they are absent.

The calculation in Fig. 3-1d, which was performed using a form of SHASTA, the first FCT algorithm developed, had an error about four or five times smaller than the simple linear methods also shown. The damping was second

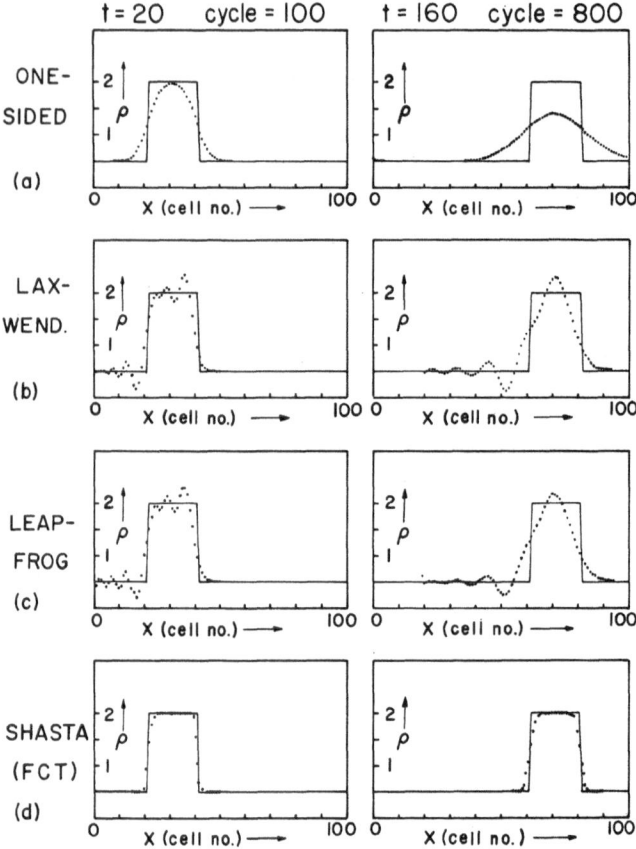

Fig. 3-1 Comparison of four difference schemes for solving the square wave problem. All tests were run with a Courant number $v\delta t/\delta x = 0.2$. The one-sided (donor cell) scheme, which is first-order accurate, is heavily diffusive. Lax–Wendroff and leapfrog are second-order schemes; they yield similar results, with phase (dispersive) errors. SHASTA, the FCT routine used, is second-order accurate except in the neighborhood of discontinuities. These can be resolved to within approximately two mesh spaces because of the nonlinear properties built into the algorithm.

order, as were the relative phase errors. The basic technique was quickly generalized to cylindrical and spherical systems, to Lagrangian as well as fixed Eulerian grids, and was applied to a number of one-, two-, and three-dimensional problems. More recent work has been directed toward extending the basic nonlinear flux-correcting techniques to convection algorithms other than SHASTA (Book, Boris and Hain, 1975) and toward discovering an "optimum" FCT algorithm (Boris and Book, 1976, 1976a).

This latter effort came to several conclusions:

1. The Gibbs error which is inherent in finite-difference techniques (Section 2.5) is two to three times smaller than the error achieved with the original FCT algorithms.
2. In local algorithms where dispersion and diffusion errors are present also, the residual dispersion errors seem to be more severe. Therefore, the optimal FCT algorithm was found to be one which has the diffusion coefficients $v_{i+1/2} = \frac{1}{6} + \frac{1}{3}\varepsilon_{i+1/2}^2$, driving dispersion errors from second to fourth order (cf. Kreiss, 1972).
3. The best of the generally useful FCT variants was almost twice as accurate as the original FCT algorithm and within 50% of the nonzero minimal error. This is roughly six to eight times more accurate than the old methods.
4. Implicit convection algorithms, in addition to being somewhat slower than the local explicit methods to compute, also have bad nonlinear properties. Information transfer occurs faster than fluid characteristic speeds, so shocks may be relatively poorly treated. Errors introduced at the boundaries propagate rapidly through the system.
5. Aside from optimizing phase errors, large diffusion/antidiffusion coefficients $v_{i+1/2} \geq \frac{1}{8}$ are necessary in order to provide enough local dissipation at shocks to satisfy the Rankine–Hugoniot conditions. This requirement is related to the one which motivated the introduction of artificial viscosity in earlier algorithms (cf. Richtmyer and Morton, 1967).

3.2 ETBFCT: A Fully Vectorized FCT Module

Since the latest FCT algorithms have eliminated 90–95% of the *removable* error and the removable error that remains is barely half of the irreducible error, it is natural to develop algorithms aimed at optimization in speed, flexibility, and generality.

We have seen that in FCT algorithms, the basic convective transport algorithm is augmented with a sufficiently strong diffusion to ensure positivity by means of a general smoothing. Since the amount of diffusion which has been added is known, FCT then performs a conservative antidiffusion step to remove the diffusion in excess of the stability limit. To preserve monotonicity, however, the antidiffusive fluxes are in effect multiplied by a coefficient which ranges from zero to unity. The criterion for choosing the reduction factors of the antidiffusive fluxes is that the antidiffused solution exhibit no new maxima or minima where the diffused solution had none. Thus positivity (or more generally, monotonicity) is built in.

Although the limit (3.3) represents the minimum amount of diffusion needed for stability, FCT algorithms generally use a larger zero-order diffusion because it has been found that the correct choice of the $\{v_{j+1/2}\}$ within the mono-

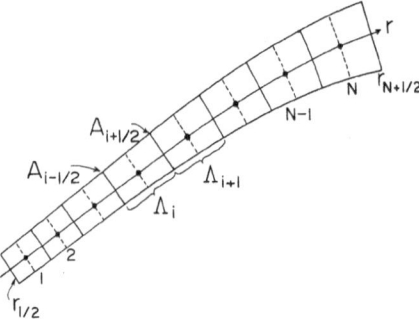

Fig. 3-2 General geometrical interpretation of the ETBFCT grid. The subroutine considers only Cartesian, cylindrical (radial), and spherical (radial) systems, but the notation also allows other geometries with a more general metric. Here r_i is the position of the ith cell, $A_{i+1/2}$ is the cross-sectional area of the interface between the ith and $(i + 1)$st cells, and Λ_i is the volume of the ith cell.

tonic and stable range will reduce convective phase errors from second to fourth order, as recommended by Kreiss (1972). Since the antidiffusion can also be chosen correspondingly larger, no real price is exacted for this improvement in phase properties. In a "phoenical" algorithm (Book, Boris, and Hain, 1975) the diffusion, transport, and antidiffusion operations are sequenced so that the algorithm reduces to the identity operator for vanishing flow velocities. The most recent efforts have taken advantage of these properties to generate minimum-operation-count FCT modules for Cartesian, cylindrical, and spherical coordinate systems with stationary (Eulerian) and movable (Lagrangian) grid systems (Boris, 1976). Below we describe the algorithm used in one of these modules, ETBFCT, and attempt to explain the reasoning behind the design decisions taken in constructing this routine. In Section 8.4 we will see what is involved in vectorizing the resulting code. Appendix A contains a listing of ETBFCT and instructions on its use.

ETBFCT (Boris, 1976) is a Fortran subroutine for solving generalized continuity equations of the form

$$\frac{\partial \rho}{\partial t} = - \frac{1}{r^{\alpha-1}} \frac{\partial}{\partial r}(r^{\alpha-1}\rho v) + \frac{1}{r^{\alpha-1}} \frac{\partial}{\partial r}(r^{\alpha-1}D_1) + C_2 \frac{\partial D_2}{\partial r} + D_3. \qquad (3.4)$$

Figure 3-2 shows a one-dimensional tube of fluid, with the velocities u^f, u^g constrained to move along the tube. The variable r measures length along the tube; u^f is the fluid flow velocity along r, i.e., the r-component of $\mathbf{v}$. The points in the center of the indicated cells are the finite-difference grid points. There are N of them and their positions at the beginning of a numerical time step are denoted by $\{r_i^o\}$ ($i = 1, \ldots, N$). At the end of a time step of duration δt the grid points are at $\{r_i^n\}$, where

$$r_i^n = r_i^o + u_i^g \delta t. \qquad (3.5)$$

Here $u_i^g \delta t$, the grid displacement, is zero for an Eulerian representation and equals $u_i^f \delta t$, the fluid displacement, in a Lagrangian mesh. Figure 3-2 also shows the basic cell volumes $\{\Lambda_i\}$ and the cell interface areas $\{A_{i+1/2}\}$. The interface areas are assumed to be perpendicular to the tube and hence to $\{u_i^f\}$. The change in the total amount of a convected quantity in a cell is just the algebraic sum of the fluxes of that quantity into and out of the cell through the interfaces. Both $\{\Lambda_i\}$ $(i = 1, \ldots, N)$ and $\{A_{i+1/2}\}$ $(i = 0, \ldots, N)$ have to be calculated using both new and old grid positions, and they have to be properly related for the representation to be fully consistent.

Figure 3-2 has N grid points and $N + 1$ interfaces, including the two at the ends of the tube. The positions of the interfaces are denoted by $\{r_{i+1/2}\}$ $(i = 0, \ldots, N)$, with superscript o or n for "old" or "new" mesh, respectively. These interface positions are calculated as

$$r_{i+1/2}^{o,n} = \tfrac{1}{2}[r_i^{o,n} + r_{i+1}^{o,n}], \qquad i = 1, \ldots, N - 1. \tag{3.6}$$

The endpoint positions, $r_{1/2}^{o,n}$ and $r_{N+1/2}^{o,n}$, have to be specified by the user. For example, they might be the location of bounding walls. Equation (3.6) could have been more general since there is no hard requirement that the interfaces be halfway between the grid points. The above choice is simple, accurate, and efficient.

We also need the flux of fluid across the interface which moves from $r_{i+1/2}^o$ to $r_{i+1/2}^n$ during the cycle. The velocities of the fluid and grid are known at the grid points, and therefore the velocity of the fluid relative to the grid is

$$\delta u_i = u_i^f - u_i^g, \qquad i = 1, \ldots, N. \tag{3.7}$$

Since the fluxes out of one cell into the next are needed on the cell interfaces, we naturally form

$$\delta u_{i+1/2} = \tfrac{1}{2}[\delta u_i + \delta u_{i+1}], \qquad i = 1, \ldots, N - 1. \tag{3.8}$$

The end values $\delta u_{1/2}$ and $\delta u_{N+1/2}$ are calculated using the specified endpoints, $r_{1/2}^{o,n}$ and $r_{N+1/2}^{o,n}$, and two endpoint velocities $u_{1/2}^f$ and $u_{N+1/2}^f$. Thus

$$\delta u_{1/2} = u_{1/2}^f - \frac{r_{1/2}^n - r_{1/2}^o}{\delta t},$$
$$\delta u_{N+1/2} = u_{N+1/2}^f - \frac{r_{N+1/2}^n - r_{N+1/2}^o}{\delta t}. \tag{3.9}$$

To complete determination of the flux we need the density at the cell interfaces. This is taken to be

$$\rho_{i+1/2}^o = \tfrac{1}{2}[\rho_{i+1}^o + \rho_i^o], \qquad i = 1, \ldots, N - 1. \tag{3.10}$$

Again we need the values on the interfaces $i = \tfrac{1}{2}$ and $i = N + \tfrac{1}{2}$. Since they are redetermined several times using updated ρ values, the formulas

$$\rho_0 = L_B \rho_1,$$
$$\rho_{N+1} = R_B \rho_N \tag{3.11}$$

are used to calculate densities one cell beyond the last cell. Then $\rho_1 - \rho_0$ and

$\rho_{N+1} - \rho_N$ are always defined at the first and last interfaces, and Eq. (3.10) gives us

$$\rho^o_{1/2} = (\tfrac{1}{2} + \tfrac{1}{2}L_B)\rho^o_1,$$
$$\rho^o_{N+1/2} = (\tfrac{1}{2} + \tfrac{1}{2}R_B)\rho^o_N. \tag{3.12}$$

Using the above definitions the convective transport part of the continuity equation can be solved:

$$\Lambda^o_i\rho^*_i = \Lambda^o_i\rho^o_i - \delta t\,\rho^o_{i+1/2}A_{i+1/2}\delta u_{i+1/2}$$
$$+ \delta t\,\rho^o_{i-1/2}A_{i-1/2}\delta u_{i-1/2}, \qquad i = 1, \ldots, N. \tag{3.13}$$

The left-hand side, $\Lambda^o_i\rho^*_i$, has not yet undergone the compression or expansion implied by the difference between the Λ^o_i and Λ^n_i. The source terms have not yet been used and the diffusion and antidiffusion portions of flux correction remain to be included.

The source terms in Eq. (3.4), if nonzero, are added into Eq. (3.13) as follows:

$$\Lambda^o_i\rho^T_i = \Lambda^o_i\rho^*_i + \tfrac{1}{2}\delta t\,A_{i+1/2}(D_{1,i+1} + D_{1,i}) - \tfrac{1}{2}\delta t\,A_{i-1/2}(D_{1,i} + D_{1,i-1})$$
$$+ \tfrac{1}{4}\delta t\,C_{2,i}(A_{i+1/2} + A_{i-1/2})(D_{2,i+1} - D_{2,i-1}) \tag{3.14}$$
$$+ \delta t\Lambda^o_i D_{3,i}, \qquad i = 2, \ldots, N-1.$$

The end values, $i = 1$ and $i = N$, are treated using D_R and D_L, the user specified right and left boundary values of D, in place of the interface average at the boundaries. ETBFCT is designed so that other source terms can be added easily to the formalism. The three source terms given provide a simple set adequate to treat many important fluid dynamics problems.

The diffusion stage of this FCT algorithm also includes the compression, calculated according to the following equation:

$$\Lambda^n_i\tilde{\rho}_i = \Lambda^o_i\rho^T_i + \nu_{i+1/2}\Lambda_{i+1/2}(\rho^o_{i+1} - \rho^o_i)$$
$$- \nu_{i-1/2}\Lambda_{i-1/2}(\rho^o_i - \rho^o_{i-1}), \qquad i = 1, \ldots, N. \tag{3.15}$$

The quantities $\{\tilde{\rho}_i\}$ correspond to the transported diffused densities $\{\rho^{td}_i\}$ defined in step 4 of the previous section. The diffusion coefficients $\{\nu_{i+1/2}\}$ which will be defined shortly, are chosen to reduce phase errors from second to fourth order. The interface volumes $\{\Lambda_{i+1/2}\}$ appear in (3.15) so that $\{\nu_{i+1/2}\}$ can be dimensionless. They are defined to be

$$\Lambda_{i+1/2} = \tfrac{1}{2}(\Lambda^n_{i+1} + \Lambda^n_i), \qquad i = 1, \ldots, N-1, \tag{3.16}$$

with the boundary interface values

$$\Lambda_{1/2} = \Lambda^n_1,$$
$$\Lambda_{N+1/2} = \Lambda^n_N. \tag{3.17}$$

The convective transport, sources, compression, and diffusion have been broken into the two stages (3.13)–(3.14) and (3.15) because we need to compute antidiffusive fluxes using $\{\rho_i^T\}$ to obtain our generalized phoenical algorithm. If we compute the antidiffusive flux using $\{\tilde{\rho}_i\}$, the algorithm will exhibit residual diffusion when the grid is Lagrangian ($u^f = u^g$) and even in the special case when both grid and fluid are stationary. The uncorrected antidiffusive fluxes are defined to be

$$F_{i+1/2} = \mu_{i+1/2}\Lambda_{i+1/2}[\rho_{i+1}^T - \rho_i^T], \qquad i = 0, \ldots, N. \tag{3.18}$$

Simple linear antidiffusion of the transported diffused solution $\{\tilde{\rho}_i\}$ would give

$$\Lambda_i^n\rho_i^n = \Lambda_i^n\tilde{\rho}_i - F_{i+1/2} + F_{i-1/2}, \qquad i = 1, \ldots, N. \tag{3.19}$$

The generalized phoenical antidiffusion is designed so that

$$\Lambda_i^n\rho_i^n = \Lambda_i^o\rho_i^o \tag{3.20}$$

when the grid is Lagrangian and $\{\delta u_{i+1/2}\}$ vanishes in Eq. (3.13). Substituting (3.13) and (3.14) into (3.15) in the Lagrangian case without sources yields

$$\Lambda_i^n\tilde{\rho}_i = \Lambda_i^o\rho_i^o + v_{i+1/2}\Lambda_{i+1/2}(\rho_{i+1}^o - \rho_i^o) - v_{i-1/2}\Lambda_{i-1/2}(\rho_i^o - \rho_{i-1}^o), \tag{3.21}$$

since $\rho_i^T = \rho_i^o$. The antidiffusion procedure of (3.18)–(3.19) used on (3.21) gives

$$\begin{aligned}\Lambda_i^n\rho_i^n = \Lambda_i^o\rho_i^o &+ (v_{i+1/2} - \mu_{i+1/2})\Lambda_{i+1/2}(\rho_{i+1}^o - \rho_i^o) \\ &- (v_{i-1/2} - \mu_{i-1/2})\Lambda_{i-1/2}(\rho_i^o - \rho_{i-1}^o).\end{aligned} \tag{3.22}$$

The desired generalized phoenical result can be achieved in (3.22) as long as $v_{i+1/2} = \mu_{i+1/2}$ whenever the grid is Lagrangian.

As explained by Boris and Book (1976), the choices

$$\begin{aligned}v_{i+1/2} &\equiv \tfrac{1}{6} + \tfrac{1}{3}\varepsilon_{i+1/2}^2, \\ \mu_{i+1/2} &\equiv \tfrac{1}{6} - \tfrac{1}{6}\varepsilon_{i+1/2}^2\end{aligned} \tag{3.23}$$

reduce the relative phase errors in a locally uniform grid region to fourth order. If we define

$$\varepsilon_{i+1/2} \equiv A_{i+1/2}\delta u_{i+1/2}\frac{\delta t}{2}\left[\frac{1}{\Lambda_i^n} + \frac{1}{\Lambda_{i+1}^n}\right], \qquad i = 0, \ldots, N, \tag{3.24}$$

the diffusion and antidiffusion coefficients are equal in the Lagrangian case, as required. Furthermore, $\{\varepsilon_{i+1/2}\}$ is computed directly from existing quantities without an extra divide because $(\Lambda_i^o)^{-1}$ and $(\Lambda_i^n)^{-1}$ are computed for use elsewhere anyway.

With $\{F_{i+1/2}\}$ from (3.18) as the raw antidiffusive flux and with $\{\tilde{\rho}_i\}$ known from (3.15), the quantities $\{S_{i+1/2}\}$ can be defined with the sign of $[\tilde{\rho}_{i+1} - \tilde{\rho}_i]$ and magnitude unity. The corrected antidiffusive flux is then calculated from the strong flux-limiting formula (Boris and Book, 1973):

$$F_{i+1/2}^c = S_{i+1/2} \max\{0, \min[|F_{i+1/2}|, S_{i+1/2}\Lambda_{i+1}^n(\tilde{\rho}_{i+2} - \tilde{\rho}_{i+1}),$$
$$S_{i+1/2}\Lambda_i^n(\tilde{\varrho}_i - \tilde{\rho}_{i-1})]\}, \quad i = 1, \ldots, N - 1. \tag{3.25}$$

For the corrected boundary fluxes $F_{1/2}^c$ and $F_{N+1/2}^c$, the min $[\ldots, \ldots, \ldots]$ term in (3.25) above contains only two terms; the correction coming from differences which are undefined beyond the boundary is effectively dropped.

The result is then computed as in (3.19), where the corrected flux $\{F_{i+1/2}^c\}$ replaces $\{F_{i+1/2}\}$. The density returned to the user is

$$\rho_i^n = \tilde{\rho}_i - \frac{1}{\Lambda_i^n}\left[F_{i+1/2}^c - F_{i-1/2}^c\right]. \tag{3.26}$$

A few of the geometric variables used above have yet to be defined. The obvious choice of volume elements in Cartesian, cylindrical, and spherical geometries is used in ETBFCT:

$$\Lambda_i^{o,n} = [r_{i+1/2}^{o,n} - r_{i-1/2}^{o,n}] \qquad \text{(Cartesian)};$$
$$\Lambda_i^{o,n} = \pi[(r_{i+1/2}^{o,n})^2 - (r_{i-1/2}^{o,n})^2] \qquad \text{(cylindrical)}; \tag{3.27}$$
$$\Lambda_i^{o,n} = \tfrac{4}{3}\pi[(r_{i+1/2}^{o,n})^3 - (r_{i-1/2}^{o,n})^3] \qquad \text{(spherical)}.$$

The corresponding interface areas are taken to be

$$A_{i+1/2} = 1 \qquad \text{(Cartesian)};$$
$$A_{i+1/2} = \pi[r_{i+1/2}^o + r_{i+1/2}^n] \qquad \text{(cylindrical)}; \tag{3.28}$$
$$A_{i+1/2} = \tfrac{4}{3}\pi[(r_{i+1/2}^o)^2 + r_{i+1/2}^o r_{i+1/2}^n + (r_{i+1/2}^n)^2] \qquad \text{(spherical)}.$$

These interface areas are time and space centered, but other centered choices are also possible. These particular definitions are forced if a constant density profile is to remain constant and unchanged when $\{u_i^f\} = 0$ and the grid is rezoned arbitrarily. If L_B and R_3 are unity, the rezone can even move fluid into and out of the system, and the density will still be constant.

The ETBFCT algorithm has been shown to generalize phoenical antidiffusion to Lagrangian systems by requiring zero residual linear diffusion whenever $u_i^f = u_i^g$. ETBFCT has also been designed to return the same constant no matter how the system is rezoned, provided that no new interfaces cross old grid point locations. Finally, in an Eulerian representation the similarity solution with constant density and linearly increasing velocity is reproduced to second order. That is, when $u_i^f = C(t)\bar{r}_i$, the continuity equation for $\rho(r,t)$ has the solution

$$\rho(r,t) = \rho(\bar{r},0) \exp\left[-\alpha \int_0^t C(t)dt\right]. \tag{3.29}$$

(Here, again, $\alpha = 1, 2, 3$ for Cartesian, cylindrical, or spherical coordinates.) Assuming C is constant (at least during one time step) and starting with

$\{\rho_i^o\} = \rho^o$(a constant), (3.13) becomes

$$\Lambda_i^o \rho_i^T = \Lambda_i^o \rho^o - \rho^o C \delta t (A_{i+1/2} \bar{r}_{i+1/2} - A_{i-1/2} \bar{r}_{i-1/2}). \tag{3.30}$$

Since the grid is stationary, $A_{i+1/2} \bar{r}_{i+1/2} - A_{i-1/2} \bar{r}_{i-1/2} = \alpha \Lambda_i^o +$ higher-order terms. Dividing (3.30) through by Λ_i^o gives

$$\rho_i^T = \rho^o (1 - \alpha C \delta t + \text{higher-order terms}). \tag{3.31}$$

In other words, the spatial constancy of ρ is preserved and ρ compresses or expands according to whether C is greater than or less than zero.

The higher-order terms can be chosen to give second-order accurate approximations to (3.29). This is done by letting $\{u_i^f\}$ be changed into the velocity which a fluid element at the grid point i would have halfway through the time step. A slight acceleration is involved since the velocity field changes in a frame of reference moving with the fluid.

Boris (1976) describes how to use ETBFCT, shows results of representative calculations, and discusses the modifications needed to run in periodic geometry and general curvilinear coordinates. He also discusses optimization on various computers. Because these algorithms, particularly the nonlinear flux-correction formula, were very carefully designed, they are fully vectorizable for pipeline and parallel processing. The execution time per continuity equation per grid point is roughly 1.3 μs on the Texas Instruments ASC at NRL. A complete time-split two-dimensional calculation on a system of 200 grid points $\times$ 200 grid points requires 2 to 2.5 s per time step, depending on the extra physics and boundary conditions incorporated in the problem.

On the CDC 7600 using the Livermore CHATR compiler the major entry ETBFCT requires 150 μs overhead plus about 25 μs per grid point per call. The Huntsville ASC executes the code at about the same speed. The ASC execution time on a one-pipe configuration for N grid points is given by the linear relation

$$t_{1 \times ASC} = 160 + 2.52N \; \mu s \tag{3.32}$$

and the timing for two pipes (the NRL configuration) is

$$t_{2 \times ASC} = 195 + 1.27N \; \mu s. \tag{3.33}$$

As an example of the type of multidimensional fluid problem which can be solved using ETBFCT, Fig. 3-3, taken from Boris (1976), shows the evolution of the Kelvin–Helmholtz instability at a shear interface. The calculation was carried out on a uniform 200 $\times$ 360 grid, starting with a weak perturbation (amplitude equal to 0.1 mesh space) and continuing through several vortex pairing times. The resolution which results permits the swirls in the Kármán vortex street to wind up through one to two revolutions before smearing out. Structures approximately three to four mesh spaces in extent are visible, and the boundaries between the two fluid regions are resolved to within approximately two mesh spaces.

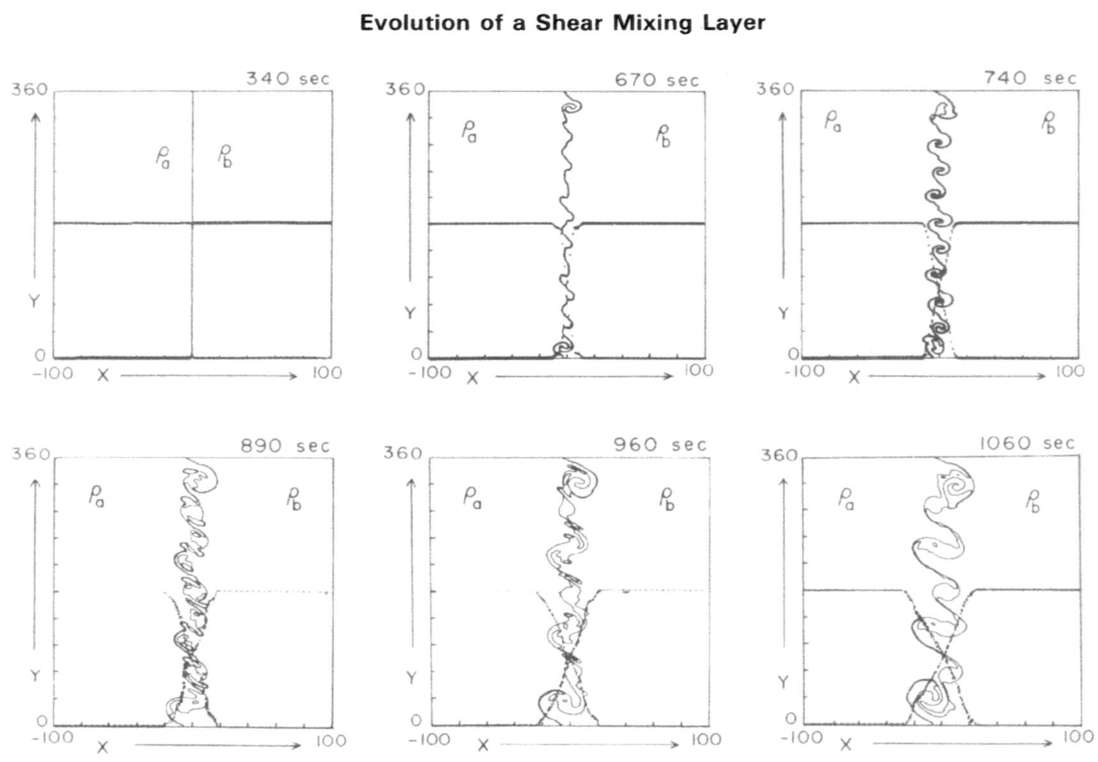

Evolution of a Shear Mixing Layer

Fig. 3-3 Sample calculations of the evolution of a Kelvin–Helmholtz instability. The time is shown in the upper right corner of each frame. The region of computation is labeled in cell units. Two equidensity regions mix as the result of a sheer-induced vortex street. Shown are contours of ρ_a/ρ, the relative density of fluid A, and density profiles for both species averaged over the y-direction, as functions of x.

3.3 Multidimensional FCT

The preceding section describes an algorithm designed to carry out multi-dimensional calculations by means of a Strang-type time-splitting procedure [see Gottlieb (1972)]. Time splitting is acceptable when it can be shown that the equations allow such a technique to be used without serious error. Such a procedure may even be preferable for programming and storage considerations. However, there are many problems for which straightforwardly applied time splitting produces unacceptable numerical results. Among these are problems involving incompressible or nearly incompressible flow fields. Here we describe a technique designed to make possible multidimensional FCT calculations without time splitting.

We consider as an example the fully two-dimensional equation

$$\rho_t + f_x + g_y = 0, \tag{3.34}$$

where ρ, f, and g are functions of x, y, and t. In finite-difference flux form we have

$$\rho_{i,j}^n = \rho_{i,j}^o - \Delta V_{i,j}^{-1}[F_{i+1/2,j} - F_{i-1/2,j} + G_{i,j+1/2} - G_{i,j-1/2}], \tag{3.35}$$

where now ρ, f, and g are defined on spatial grid points x_i, y_j at time levels t^o, t^n, and ΔV_{ij} is a two-dimensional area element centered on grid point (i,j). Now there are two sets of transportive fluxes F and G, and the FCT algorithm proceeds as before:

(1) Compute $F_{i+1/2,j}^L$ and $G_{i,j+1/2}^L$ by a low-order monotonic scheme.
(2) Compute $F_{i+1/2,j}^H$ and $G_{i,j+1/2}^H$ by a high-order scheme.
(3) Define the antidiffusive fluxes:

$$A_{i+1/2,j} \equiv F_{i+1/2,j}^H - F_{i+1/2,j}^L;$$
$$A_{i,j+1/2} \equiv G_{i,j+1/2}^H - G_{i,j+1/2}^L.$$

(4) Compute the low-order time-advanced solution:

$$\rho_{i,j}^{td} = \rho_{i,j}^o - \Delta V_{i,j}^{-1}[F_{i+1/2,j}^L - F_{i-1/2,j}^L + G_{i,j+1/2}^L - G_{i,j-1/2}^L].$$

(5) Limit the antidiffusive fluxes:

$$A_{i+1/2,j}^C = A_{i+1/2,j}\, C_{i+1/2,j}, \qquad 0 \le C_{i+1/2,j} \le 1;$$
$$A_{i,j+1/2}^C = A_{i,j+1/2}\, C_{i,j+1/2}, \qquad 0 \le C_{i,j+1/2} \le 1.$$

(6) Apply the limited antidiffusive fluxes:

$$\rho_{i,j}^n = \rho_{i,j}^{td} - \Delta V_{i,j}^{-1}[A_{i+1/2,j}^C - A_{i-1/2,j}^C + A_{i,j+1/2}^C - A_{i,j-1/2}^C].$$

As can be easily seen, implementation of FCT in multidimensions is straightforward, with the exception of step 5.

We describe now in one spatial dimension a flux-limiting algorithm (Zalesak, 1979) which generalizes easily to multidimensions and which, even in one dimension, exhibits greater flexibility with respect to peaked profiles than the limiter described by Boris and Book (1973).

We seek to limit the antidiffusive flux $A_{i+1/2}$ such that

$$A^C_{i+1/2} = C_{i+1/2} A_{i+1/2}, \qquad 0 \leq C_{i+1/2} \leq 1, \tag{3.36}$$

and such that $A^C_{i+1/2}$ acting in concert with $A^C_{i-1/2}$ will not allow

$$\rho^n_i = \rho^{td}_i - \Delta x_i^{-1} [A^C_{i+1/2} - A^C_{i-1/2}] \tag{3.37}$$

to exceed some maximum value ρ^{max}_i nor fall below some minimum value ρ^{min}_i. We leave the determination of ρ^{max}_i and ρ^{min}_i until later.

We define three quantities:

P^+_i = the sum of all antidiffusive fluxes into grid point i

$$= \max(0, A_{i-1/2}) - \min(0, A_{i+1/2}). \tag{3.38}$$

$$Q^+_i = (\rho^{max}_i - \rho^{td}_i) \Delta x_i. \tag{3.39}$$

$$R^+_i = \begin{cases} \min(1, Q^+_i/P^+_i), & P^+_i > 0; \\ 0, & P^+_i = 0. \end{cases} \tag{3.40}$$

Provided that $\rho^{max}_i \geq \rho^{td}_i$ (which must hold), all three of the above quantities are positive, and R^+_i represents the least upper bound on the fraction which must multiply all antidiffusive fluxes into grid point i to guarantee no overshoot at grid point i.

Similarly we define three corresponding quantities:

P^-_i = the sum of all antidiffusive fluxes away from grid point i

$$= \max(0, A_{i+1/2}) - \min(0, A_{i-1/2}). \tag{3.41}$$

$$Q^-_i = (\rho^{td}_i - \rho^{min}_i) \Delta x_i. \tag{3.42}$$

$$R^-_i = \begin{cases} \min(1, Q^-_i/P^-_i), & P^-_i > 0; \\ 0, & P^-_i = 0. \end{cases} \tag{3.43}$$

Again assuming that $\rho^{min}_i \leq \rho^{td}_i$, we find that R^-_i represents that least upper bound on the fraction which must multiply all antidiffusive fluxes away from grid point i to guarantee that there be no undershoot at grid point i.

Finally we observe that all antidiffusive fluxes are directed away from one grid point and into an adjacent one. Limiting will therefore take place with respect to undershoots for the former and with respect to overshoots for the latter. A guarantee that neither event come to pass demands our taking a minimum:

$$C_{i+1/2} = \begin{cases} \min(R^+_{i+1}, R^-_i), & A_{i+1/2} \geq 0; \\ \min(R^+_i, R^-_{i+1}), & A_{i+1/2} < 0. \end{cases} \tag{3.44}$$

Based on practical experience with the strong flux limiter (3.25), we set

$$A_{i+1/2} = 0 \tag{3.45}$$

if simultaneously

$$A_{i+1/2}(\rho_{i+1}^{td} - \rho_i^{td}) < 0, \tag{3.46}$$

and either

$$A_{i+1/2}(\rho_{i+2}^{td} - \rho_{i+1}^{td}) < 0, \tag{3.47}$$

or

$$A_{i+1/2}(\rho_i^{td} - \rho_{i-1}^{td}) < 0. \tag{3.48}$$

In practice, the effect of (3.45) is minimal and is primarily cosmetic in nature. This is because cases of antidiffusive fluxes directed down gradients in ρ^{td} are rare, and even when they occur they usually involve flux magnitudes that are small compared to adjacent fluxes. If the adjustment (3.45)–(3.48) is used, it should be applied before (3.36)–(3.44).

We come now to a determination of the quantities ρ_i^{max} and ρ_i^{min} in (3.39) and (3.42). A safe choice is

$$\rho_i^{max} = \max(\rho_{i-1}^{td}, \rho_i^{td}, \rho_{i+1}^{td}), \tag{3.49}$$

$$\rho_i^{min} = \min(\rho_{i-1}^{td}, \rho_i^{td}, \rho_{i+1}^{td}). \tag{3.50}$$

This choice will produce results identical with those of the Boris–Book (1973) formulation of step 5 in one dimension, including the occurrence of the "clipping" phenomenon to be mentioned shortly.

A better choice is given by

$$\rho_i^a = \max(\rho_i^o, \rho_i^{td}); \tag{3.51}$$

$$\rho_i^{max} = \max(\rho_{i-1}^a, \rho_i^a, \rho_{i+1}^a). \tag{3.52}$$

$$\rho_i^b = \min(\rho_i^o, \rho_i^{td}); \tag{3.53}$$

$$\rho_i^{min} = \min(\rho_{i-1}^b, \rho_i^b, \rho_{i+1}^b). \tag{3.54}$$

This choice allows us to look back to the previous time step for upper and lower bounds on ρ_i^n. It is clear that these two methods of determining ρ_i^{max} and ρ_i^{min} represent only a small sample of possible methods. The alternative flux limiter described in (3.36)–(3.48) admits any physically motivated upper and lower bound on ρ_i^n supplied by the user, introducing a flexibility unavailable with the original flux limiter. The alternative flux-limiting algorithm just presented generalizes trivially to any number of dimensions. For the sake of completeness we present here the algorithm for two spatial dimensions.

We seek to limit the antidiffusive fluxes $A_{i+1/2,j}$ and $A_{i,j+1/2}$ so that

$$A^C_{i+1/2,j} = C_{i+1/2,j}A_{i+1/2,j}, \qquad 0 \le C_{i+1/2,j} \le 1, \tag{3.55}$$

$$A^C_{i,j+1/2} = C_{i,j+1/2}A_{i,j+1/2}, \qquad 0 \le C_{i,j+1/2} \le 1, \tag{3.56}$$

and so that $A^C_{i+1/2,j}$, $A^C_{i,j+1/2}$, $A^C_{i-1/2,j}$ and $A^C_{i,j-1/2}$ acting in concert will not cause

$$\rho^n_{i,j} = \rho^{td}_{i,j} - \Delta V^{-1}_{i,j}[A^C_{i+1/2,j} - A^C_{i-1/2,j} + A^C_{i,j+1/2} - A^C_{i,j-1/2}] \tag{3.57}$$

to exceed some maximum value $\rho^{max}_{i,j}$ or to fall below some minimum value $\rho^{min}_{i,j}$.

Again we compute six quantities completely analogous to those computed in (3.38)–(3.43):

$P^+_{i,j} =$ the sum of all antidiffusive fluxes into grid point (i,j)

$$= \max(0,A_{i-1/2,j}) - \min(0,A_{i+1/2,j}) + \max(0,A_{i,j-1/2}) \tag{3.58}$$
$$- \min(0,A_{i,j+1/2}).$$

$$Q^+_{i,j} = (\rho^{max}_{i,j} - \rho^{td}_{i,j})\Delta V_{i,j}. \tag{3.59}$$

$$R^+_{ij} = \begin{cases} \min(1,Q^+_{ij}/P^+_{ij}), & P^+_{ij} > 0; \\ 0, & P^+_{ij} = 0. \end{cases} \tag{3.60}$$

$P^-_{i,j} =$ the sum of all antidiffusive fluxes away from grid point (i,j)

$$= \max(0,A_{i+1/2,j}) - \min(0,A_{i-1/2,j}) + \max(0,A_{i,j+1/2}) \tag{3.61}$$
$$- \min(0,A_{i,j-1/2}).$$

$$Q^-_{i,j} = (\rho^{td}_{i,j} - \rho^{min}_{i,j})\Delta V_{i,j}. \tag{3.62}$$

$$R^-_{i,j} = \begin{cases} \min(1,Q^-_{i,j}/P^-_{i,j}), & P^-_{i,j} > 0; \\ 0, & P^-_{i,j} = 0. \end{cases} \tag{3.63}$$

Equation (3.44) becomes

$$C_{i+1/2,j} = \begin{cases} \min(R^-_{i+1,j},R^+_{i,j}), & A_{i+1/2,j} < 0; \\ \min(R^-_{i,j},R^+_{i+1,j}), & A_{i+1/2,j} \ge 0. \end{cases} \tag{3.64}$$

$$C_{i,j+1/2} = \begin{cases} \min(R^-_{i,j+1},R^+_{i,j}), & A_{i,j+1/2} < 0; \\ \min(R^-_{i,j},R^+_{i,j+1}), & A_{i,j+1/2} \ge 0. \end{cases} \tag{3.65}$$

Equations (3.45)–(3.48) become

$$A_{i+1/2,j} = 0 \tag{3.66}$$

if simultaneously

$$A_{i+1/2,j}(\rho^{td}_{i+1,j} - \rho^{td}_{i,j}) < 0, \tag{3.67}$$

and either

$$A_{i+1/2,j}(\rho^{td}_{i+2,j} - \rho^{td}_{i+1,j}) < 0 \tag{3.68}$$

or

$$A_{i+1/2,j}(\rho_{i,j}^{td} - \rho_{i-1,j}^{td}) < 0. \tag{3.69}$$

Likewise,

$$A_{i,j+1/2} = 0 \tag{3.70}$$

if simultaneously

$$A_{i,j+1/2}(\rho_{i,j+1}^{td} - \rho_{i,j}^{td}) < 0, \tag{3.71}$$

and either

$$A_{i,j+1/2}(\rho_{i,j+2}^{td} - \rho_{i,j+1}^{td}) < 0 \tag{3.72}$$

or

$$A_{i,j+1/2}(\rho_{i,j}^{td} - \rho_{i,j-1}^{td}) < 0. \tag{3.73}$$

Equations (3.51) and (3.54) become

$$\rho_{i,j}^a = \max(\rho_{i,j}^o, \rho_{i,j}^{td}), \tag{3.74}$$

$$\rho_{i,j}^{max} = \max(\rho_{i-1,j}^a, \rho_{i,j}^a, \rho_{i+1,j}^a, \rho_{i,j-1}^a, \rho_{i,j+1}^a); \tag{3.75}$$

$$\rho_{i,j}^b = \min(\rho_{i,j}^o, \rho_{i,j}^{td}), \tag{3.76}$$

$$\rho_{i,j}^{min} = \min(\rho_{i-1,j}^b, \rho_{i,j}^b, \rho_{i+1,j}^b, \rho_{i,j-1}^b, \rho_{i,j+1}^b). \tag{3.77}$$

Again, the effect of (3.66)–(3.73) is minimal, but if this refinement is used, it should be applied before (3.55)–(3.65). Note that our search for $\rho_{i,j}^{max}$ and $\rho_{i,j}^{min}$ now extends over both coordinate directions. Where finite gradients exist in both directions, this procedure will allow us to stop the clipping phenomenon in regions where a peak exists with respect to one coordinate direction but not the other. The subroutine FLIMIT listed in Appendix B allows the user a choice of whether to use strong flux limiting or the formulation (3.58)–(3.77).

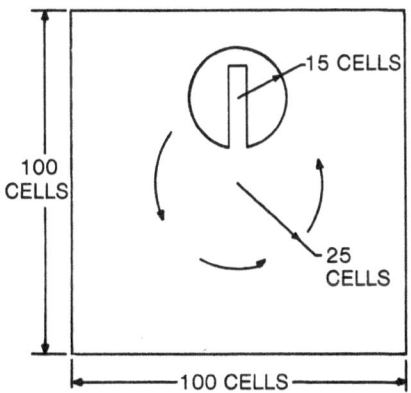

Fig. 3-4 Schematic representation of two-dimensional solid body rotation problem. Initially $\rho = 3.0$ inside the cutout cylinder while outside $\rho = 1.0$. The rotational speed is such that one full revolution is effected in 628 cycles. The width of the gap separating the two halves of the cylinder, as well as the maximum extent of the "bridge" connecting the two halves, is 5 cells.

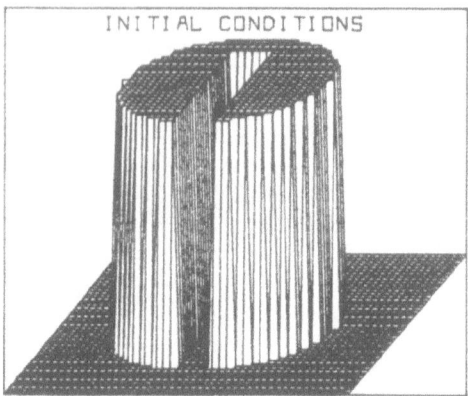

Fig. 3-5 Perspective view of
initial conditions for the two-
dimensional solid body rotation
problem. Note that only the
50 × 50 portion of the mesh
centered on the cylinder is
displayed.

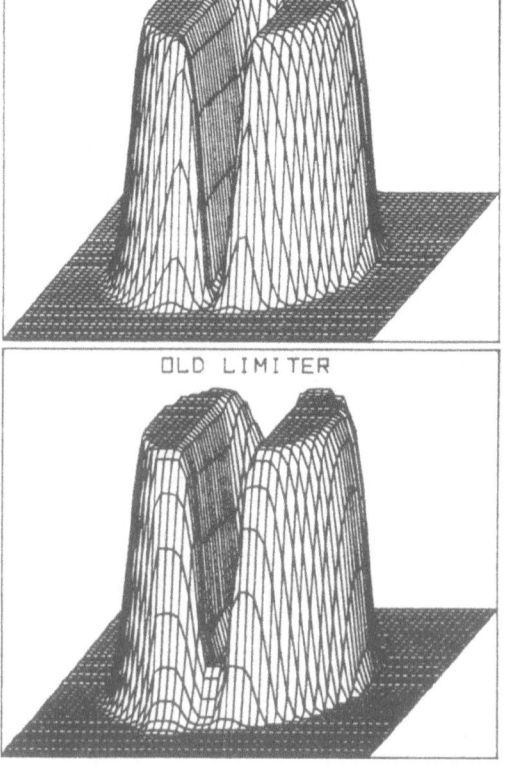

Fig. 3-6 Comparison of
perspective views of the ρ
profile after 157 iterations
(one-quarter revolution) with
both the strong ("old") and
generalized ("new") flux
limiters. The perspective view
has been rotated with the
cylinder, so that direct
comparison with Fig. 3-5 can
be made. Again we plot only
the 50 × 50 grid centered on
the analytic center of the
cylinder. Features to compare
are the filling in of the gap,
erosion of the "bridge," and
the relative sharpness of the
profiles defining the front
surface of the cylinder.

As a test problem, consider solid body rotation (cf. Forester, 1977). That is, we have (3.34) with $f = \rho v_x$, $g = \rho v_y$, $v_x = -\Omega(y - y_0)$, and $v_y = \Omega(x - x_0)$. Here Ω is the (constant) angular velocity in rad/second and (x_0, y_0) is the position of the axis of rotation. The configuration is shown in Fig. 3-4. The computational grid is 100×100 cells, $\Delta x = \Delta y$, with counter-clockwise rotation taking place about grid point (50,50). Centered at grid point (50,75) is a cylinder of radius 15 grid points, through which a slot has been cut of width 5 grid points. The time step and rotational speed are chosen such that 628 time steps will effect one complete revolution of the cylinder about the central point. A perspective view of the initial conditions is shown in Fig. 3-5.

Our high-order scheme for the following tests is a fully two-dimensional, fourth-order in space, second-order in time leapfrog–trapezoidal scheme, the leapfrog step of which is a two-dimensional fourth-order Kreiss–Oliger (1972) scheme. The low-order scheme is simply two-dimensional donor cell plus

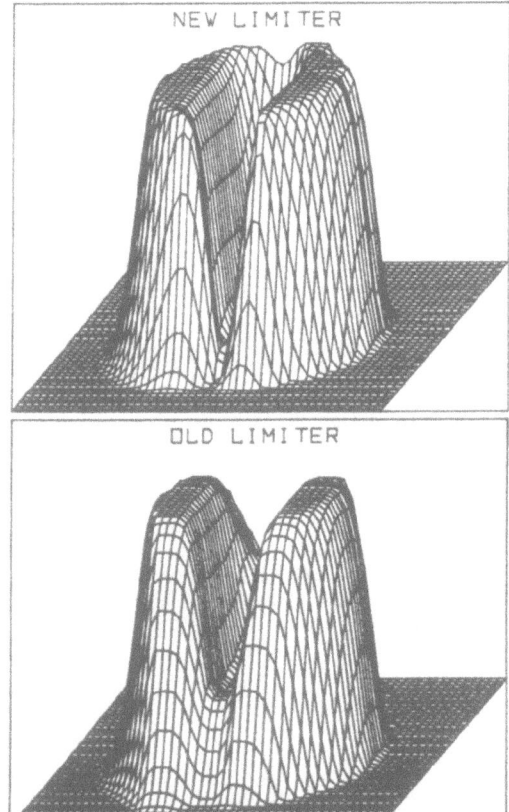

Fig. 3-7 Same as Fig. 3-6, but after 628 iterations (one full revolution). Note decreased diffusion with new flux limiter.

a two-dimensional zeroth-order diffusion term with diffusion coefficient $\frac{1}{8}$.

In Fig. 3-6 we show a perspective view of the two calculations after one-quarter revolution (157 iterations). Figure 3-7 presents a comparison of the results of the two calculations for one full revolution (628 cycles). Two features are obvious. The first is a much greater filling-in of the slot with the time split than with the fully two-dimensional flux limiter. The second is the loss of the bridge connecting the two halves of the cylinder in the case of the time-split approximation to (3.34). Less obvious is the lack of clipping of the peaked profiles defining the front surface of the cylinder with the fully multidimensional limiter. Clearly this is due to the fact that the multidimensional flux limiter can look in both directions to determine whether or not a genuine maximum exists. Note that there are two factors working in favor of the fully multidimensional flux limiter: (1) this ability to look in both directions to find minima and maxima; and (2) the ability to scan both $\rho^o_{i,j}$ and $\rho^{td}_{i,j}$ to find maxima and minima. Both of these factors are responsible for the improved profiles.

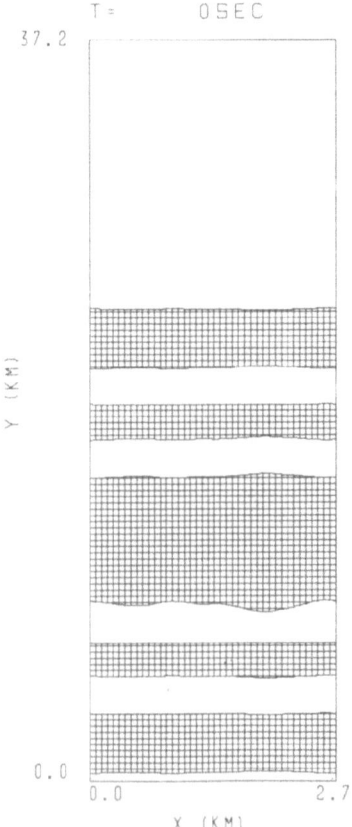

T =   0 SEC

37.2

Y (KM)

0.0

0.0   2.7

X (KM)

Fig. 3-8 Isodensity contours of plasma density at $t = 0$. The initial distribution for n_e/n_o is a Gaussian in y, centered at $y = 12.1$ km, plus a small random perturbation in x. Contours are drawn for $n_e/n_o = 1.5, 3.5, 5.5, 7.5,$ and 9.5. The area between every other contour line is cross-hatched. Only 120 of the 160 cells actually used in the y-direction are displayed. Boundary conditions are periodic in both directions. In our plot $\mathbf{B}_0$ is toward the reader, and $\mathbf{E}_0$ is directed toward the right, and we have placed ourselves in a frame moving with velocity $(c/B_0^2)\mathbf{E}_0 \times \mathbf{B}_0$. The upper portion of the Gaussian is physically unstable to perturbations, while the lower half is (linearly) stable.

A physical problem treated by this technique which is mathematically close to that of the vorticity dynamics formulation (2.7)–(2.9) is that of barium cloud striations. A two-dimensional $(x–y)$ plasma cloud initialized in a region of constant magnetic field $\mathbf{B}_0$ directed along the z-axis, with an externally imposed electric field $\mathbf{E}_0$ directed along the x-axis, will tend to drift in the $\mathbf{E}_0 \times \mathbf{B}_0$ direction (along the negative y-axis). If the ion–neutral collision frequency is finite, Pedersen conductivity effects will produce polarization fields which tend to shield the inner (more dense) regions of the cloud from $\mathbf{E}_0$, causing this inner portion of the cloud to drift more slowly than the outer portions of the cloud. This results in a steepening of gradients on the back side of the cloud. Arguments similar to those above, applied to infinitesimal perturbations imposed upon this back side gradient, show that the back side of the cloud is physically unstable to perturbations along $\hat{x}$. For a detailed description of this problem, see Scannapieco *et al.* (1976).

Figures 3-8 to 3-12 show isodensity contours of reduced electron density n_e/n_0 for the above configuration at various times in the integration. It is seen

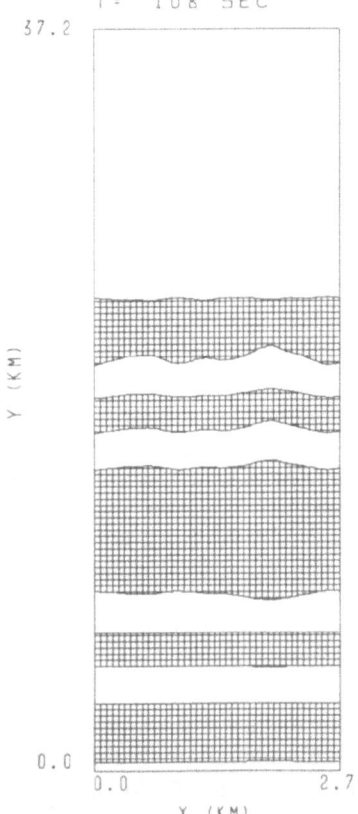

Fig. 3-9 Same as Fig. 3-8, but for $t = 108$ s. Note slow linear growth on unstable side.

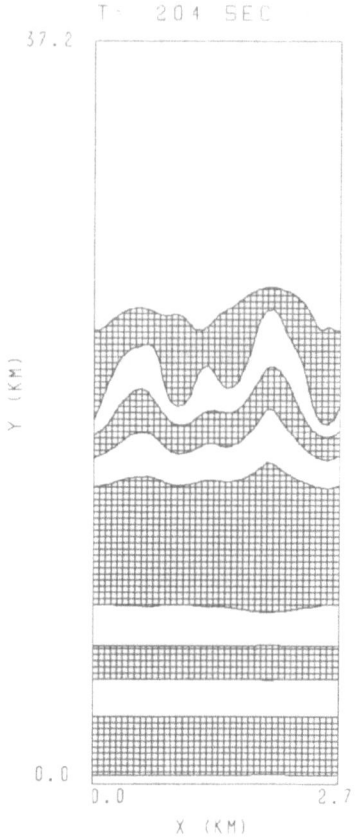

Fig. 3-10 Same as Fig. 3-8, but for $t = 204$ s. Growth is now much more rapid, and the calculation is entering a highly nonlinear regime.

that, as expected, the back of the cloud (the upper half in the plots) is unstable, growing linearly in the very early stages of development. Nonlinear effects soon enter the physics, however, as each striation successively bifurcates, producing smaller and smaller scale structures, in agreement with observation of the results of the ionospheric barium cloud releases which we are attempting to model. Two points which bear on the numerics should be noted: (1) the intense gradients dictated by the physics are *not* diffused away, nor do there appear in the problem any of the "ripples" associated with numerical dispersion which normally appear when steep gradients try to form; (2) precisely because we did not have to resort to time splitting, none of the usual time-splitting phenomena, such as temporal density oscillations and spurious density values, are evident.

On NRL's Texas Instruments ASC computer, the test problem calculations required 93 and 125 s of CPU time for the time-split and fully multidimensional cases, respectively, a cost penalty of slightly more than 30% for the multi-dimensional limiter. Of course, this extra cost is highly problem dependent.

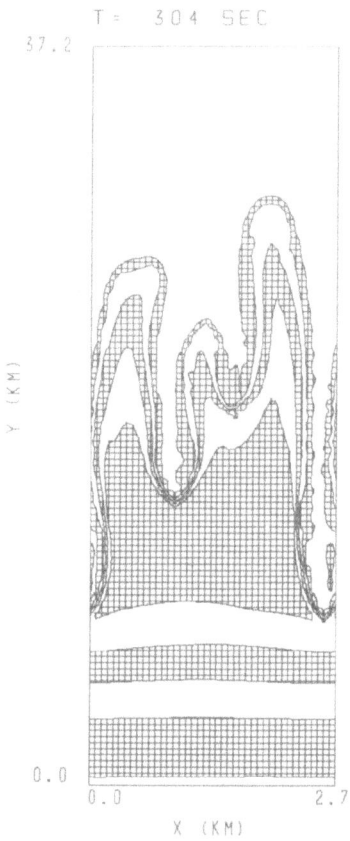

T = 304 SEC

Fig. 3-11 Same as Fig. 3-8, but for $t = 304$ s. Development is fully nonlinear, as the intense gradients and associated high Fourier wave numbers become apparent.

For instance, the striations code spends 80% of its time solving the Poisson-like elliptic equation, making the net cost penalty of fully multidimensional flux limiting only a few percent.

As a final example, we consider planar constant-velocity ideal-gas shocks reflecting from wedges. Under certain circumstances Mach stems are formed. Historically these configurations have proven to be exceedingly difficult to calculate with high accuracy. For unreacting flows at Mach numbers greater than about 2.5 and wedge angles between 20° and 50°, double Mach stems develop. Numerical schemes previously used for this problem reproduce qualitatively the wave structure and shape, but have some difficulty making accurate predictions of flow details such as density contours (a conclusion drawn by Ben-Dor and Glass, 1978) even in the single-Mach-stem case.

We believe that the calculational difficulties experienced with this problem were the result of excessive numerical diffusion, especially in the region of the contact surface. Book et al. (1980) have successfully performed numerical simulations for various shock strengths and wedge angles, using both a time-

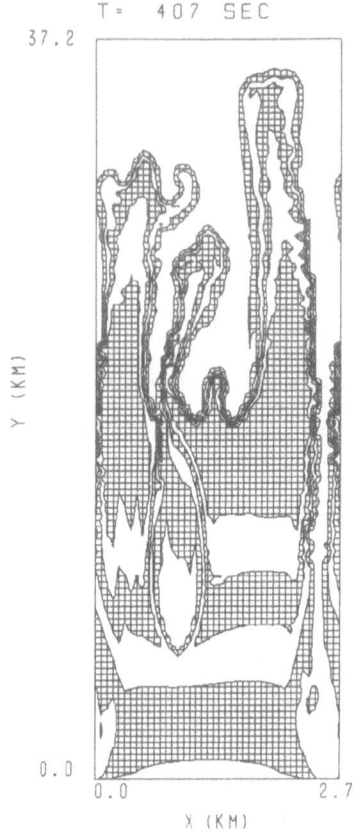

Fig. 3-12 Same as Fig. 3-8, but for $t = 407$ s. Several plasma bifurcations are apparent, in agreement with the experimental results from ionospheric barium cloud releases, and maximum-to-minimum density variations are resolved over only two cells.

split algorithm similar to ETBFCT and the fully two-dimensional algorithm described in the present section.

Examples of the calculated density contours and wave structure for the complex and double-Mach-reflection cases obtained using the multidimensional form of FCT are shown in Fig. 3-13. The incident shock, I, the contact surface, CS, and the first and second Mach stems, M_1 and M_2, are indicated. Note in particular the forward curl of the contact surface near the wall and the small region (4 × 7 mesh points) of high-density gas just to the left of the point where the contact surface reaches the wall. The latter causes a second peak in the pressure and density distribution on the wall, as shown in Fig. 3-14. The accuracy of the calculations has been verified by comparison with experimental density distributions along the wall, as shown in Fig. 3-15, and with experimental pressure measurements. FCT provides adequate resolution of the key surfaces (contact surface and second Mach stem) in regions as small as 5 × 5 cells.

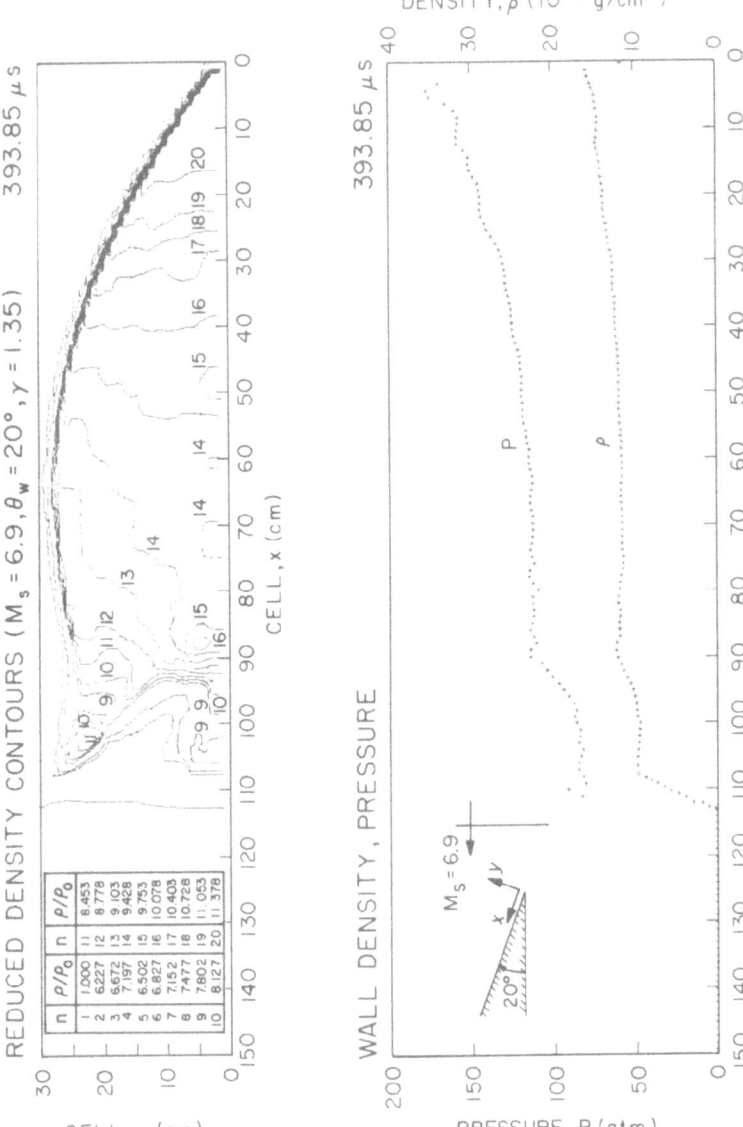

Fig. 3-13 Wave structure and density contours for complex and double Mach reflection from a wedge. The reflecting condition is imposed at the bottom of the mesh, defining the wedge surface. The other boundary conditions impose the values of the fluid variables ahead of or behind the incident shock, depending on the location of the shock front as a function of time. The adiabatic index is taken to be a constant, $\gamma = 1.35$.

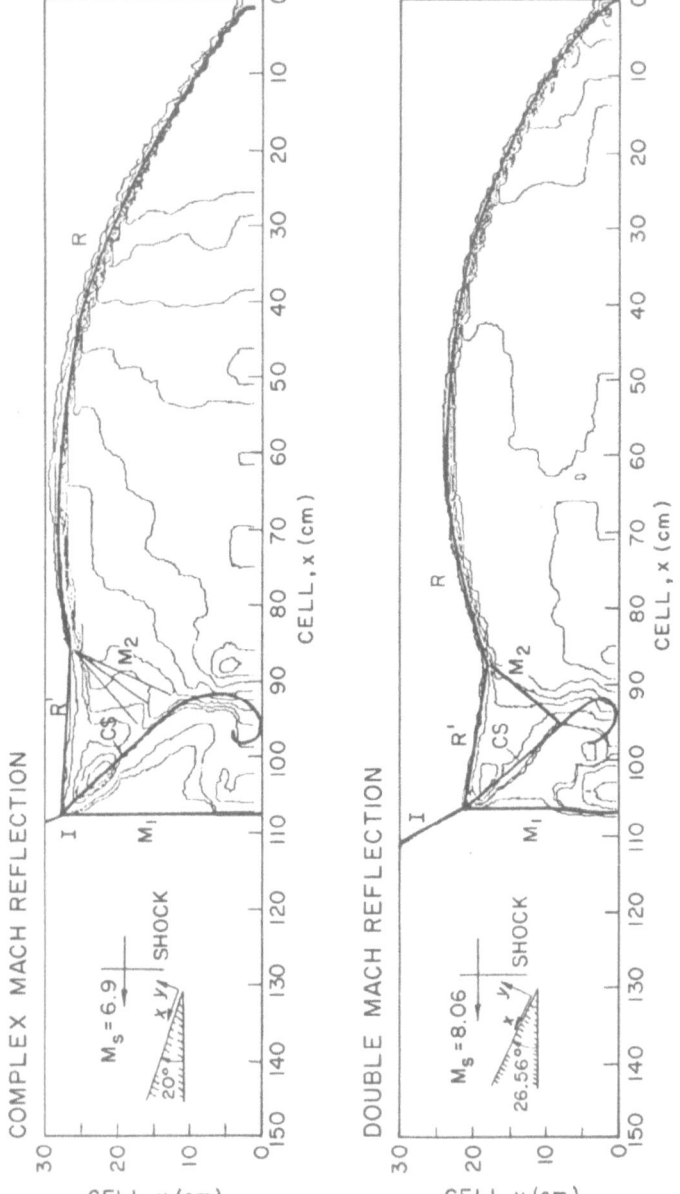

Fig. 3-14 (a) Density contours for the complex Mach reflection case of Fig. 3-13 ($M = 6.9$, $\theta_w = 20°$, $\gamma = 1.35$), chosen to agree with those of Ben-Dor and Glass (1978). (b) Corresponding pressure loading and density on the wedge, plotted against cell number.

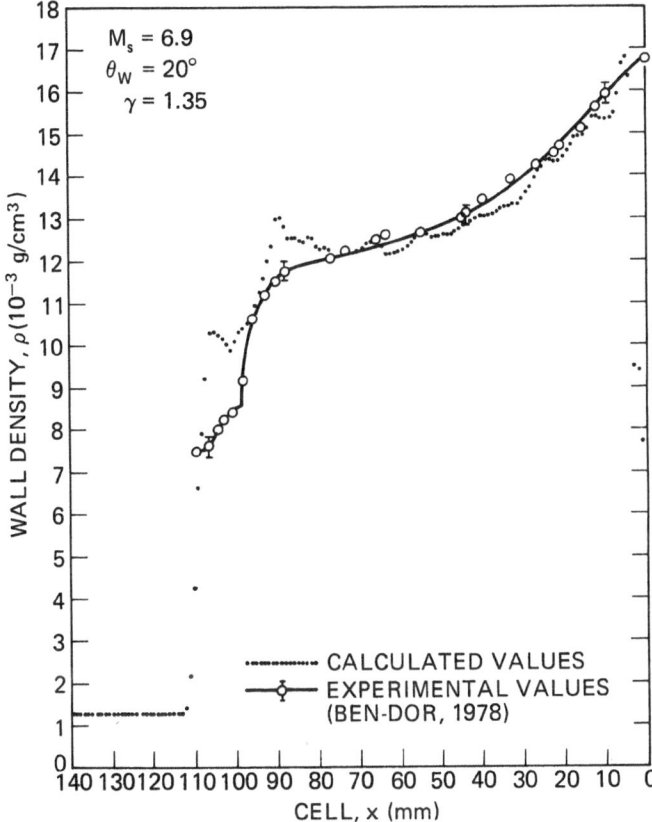

Fig. 3-15 Comparison of the calculated density profile on the wedge (Fig. 3-14b) with measured values (Ben-Dor and Glass, 1978). Agreement is good, except at the location of the Mach stem, where experimental limits on resolution and real air effects (which tend to reduce the effective value of γ) cause the measured density to be lower than that calculated.

CHAPTER 4

Efficient Time Integration Schemes for Atmosphere and Ocean Models

4.1 Introduction

As atmospheric numerical models increased in complexity, the need for computationally efficient time integration schemes became apparent. Early efforts along this line began with Marchuk (1974) and Robert *et al.* (1972). They noted that atmospheric motions can be separated into two classes; the Rossby modes, where nonlinearity and quasigeostrophic balance (i.e., balance between Coriolis and pressure gradient forces) play an important role, and gravity modes, for which the pressure gradient terms are balanced by the inertial terms. It was also observed that the high-frequency gravity modes, which propagate faster than the Rossby modes, are quasilinear and carry only a small portion of the total energy.

The barotropic or Lamb mode (the surface gravity mode) is the fastest moving mode allowed in primitive equation models. It travels with a phase speed of about 300 ms^{-1}, about an order of magnitude faster than the fastest Rossby mode. Even though the fast-moving surface gravity mode contains negligibly small fractions of the total energy, it determines the integration time step for explicit integration of the primitive equations. The Rossby modes which contain most of the atmospheric energy have phase speeds at least a factor of 4 slower than the external gravity modes. Explicit methods are therefore very inefficient for these modes.

Assuming that the high-frequency gravity waves are quasilinear, Robert *et al.* (1972) developed a technique which treated the terms governing these modes, called the semi-implicit method. This technique allows the use of time steps much longer than those allowed by the surface gravity waves. Now the time step is determined by the Rossby modes. This increase in step has the physical effect of retarding those gravity modes for which the time step violates the CFL condition. The Rossby modes are still treated accurately, as the time step is consistent with the CFL condition for these.

The major penalty incurred in this method is that the implicit treatment of the gravity waves requires the solution of a two-dimensional Helmholtz equation for each natural gravity eigenmode of the model. Since all the eigenmodes are treated implicitly, the semi-implicit method requires that as many Helmholtz equations be solved as there are eigenmodes in the model.

In general, the solution of these equations is time-consuming, which offsets the advantage gained through increasing the time step.

Since less than a third of the gravity eigenmodes of the atmosphere travel faster than the Rossby modes, Burridge (1975) developed a method which treats implicitly only those that travel faster, while the slower modes are treated explicitly. This significantly reduces the number of elliptic equations which need to be solved, thus improving the efficiency of the method. Unfortunately, the Burridge formulation suffers from time truncation errors which vary depending on the order of approximation used for different terms. These errors restrict the time step to about half of that allowed in semi-implicit models and make it less efficient than these methods.

As shown by Kreiss and Oliger (1972), the implicit treatment retards the modes for which the CFL condition is violated more than necessary for stability. As a consequence, if the time step is too close to the CFL condition of the Rossby modes, the geostrophic adjustment time of the Rossby modes may be affected. This means that the gravity waves which should restore the geostrophic balance propagate too slowly, causing serious errors in the small scale structure of the fast-moving Rossby modes.

Realizing the problems and limitations associated with the implicit methods, Gadd (1978) integrated both the gravity wave contribution and Rossby wave contribution explicitly using two time steps: a smaller time step for gravity modes and a larger one for the Rossby modes. Since the linear terms require a small fraction of the operations required by the other terms, this increases efficiency by about a factor of 3 over explicit methods. However, the scheme also suffers from variable time truncation errors, and some of the important conservative properties of the differential equations cannot be maintained by this method. This reduces the accuracy of the solution for long integration times.

This chapter describes a new technique which is related to the split methods developed by Burridge (1975) and Gadd (1978). Even though it has the same name as the latter (both are called split-explicit methods), it is substantially different. In Gadd's method all the gravity wave contributions are integrated with the same time step. In the split-explicit method, we separate the terms into those governing the gravity modes and those governing the Rossby modes. The split equations are then integrated with their time steps given by the respective CFL conditions. The various terms contribute additively instead of multiplicatively as in ordinary time-splitting methods. This provides a more accurate solution than the other two methods.

The split-explicit method is described in Sections 4.2 and 4.3 for barotropic and baroclinic models, respectively. Section 4.4 describes the extension of the method to ocean models.

4.2 Time Integration Schemes for Barotropic Models

Definition of three integration schemes for barotropic models: The governing equations for flow in a barotropic fluid, i.e., one in which pressure is a function of density, are

$$\frac{\partial u}{\partial t} + \frac{\partial \phi}{\partial x} = A_u, \tag{4.1}$$

$$\frac{\partial v}{\partial t} + \frac{\partial \phi}{\partial y} = A_v, \tag{4.2}$$

$$\frac{\partial \phi}{\partial t} + \Phi \left(\frac{\partial u}{\partial x} + \frac{\partial v}{\partial y} \right) = A_\phi, \tag{4.3}$$

where u and v are the horizontal components of the velocity, and Φ and ϕ represent the mean and perturbed geopotentials, respectively. Inertial, non-linear, frictional, and source terms are represented by the terms A_u, A_v, and A_ϕ. Equations (4.1)–(4.3) constitute the shallow-water formulation of the equations for incompressible three-dimensional flow.

Equations (4.1)–(4.3) can be written in finite-difference form in space by replacing the spatial derivatives by second-order finite differences over a staggered grid (Fig. 4-1) as follows:

$$\frac{\partial u}{\partial t} + \delta_x \phi = A_u, \tag{4.4}$$

$$\frac{\partial v}{\partial t} + \delta_y \phi = A_v, \tag{4.5}$$

$$\frac{\partial \phi}{\partial t} + \Phi D = A_\phi, \tag{4.6}$$

where the difference operation δ is defined as

$$\delta_\alpha \beta = \frac{\beta(\alpha + \frac{1}{2}\Delta\alpha) - \beta(\alpha - \frac{1}{2}\Delta\alpha)}{\Delta\alpha}. \tag{4.7}$$

Here α and β represent any independent and dependent variables, respectively. The variable $D = \delta_x u + \delta_y v$ represents the finite-difference form of the velocity divergence. For notational convenience the right-side terms (RST) in Eqs. (4.1)–(4.3) and (4.4)–(4.6) are represented by the same variables. It should be noted, however, that the RST in (4.4)–(4.6) are the finite-difference form of the corresponding terms in (4.1)–(4.3).

Integrating Eqs. (4.4)–(4.6) from $t - \Delta t$ to $t + \Delta t$, we obtain

$$u(t + \Delta t) - u(t - \Delta t) + 2\Delta t \delta_x \bar{\phi} = 2\Delta t \bar{A}_u, \tag{4.8}$$

$$v(t + \Delta t) - v(t - \Delta t) + 2\Delta t \delta_y \bar{\phi} = 2\Delta t \bar{A}_v, \tag{4.9}$$

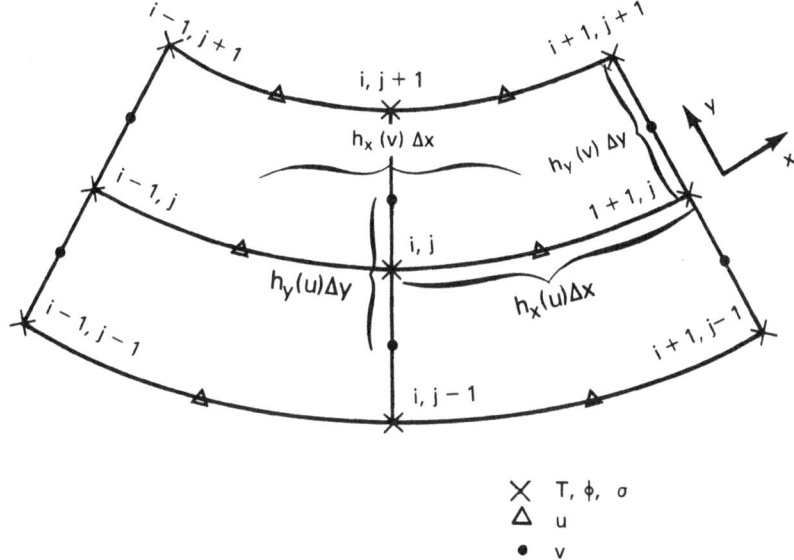

Fig. 4-1 Staggered grid system in general curvilinear horizontal coordinates. Here h_x and h_y are general scale (map) factors.

$$\phi(t + \Delta t) - \phi(t - \Delta t) + 2\Delta t \Phi \bar{D} = 2\Delta t \bar{A}_\phi, \qquad (4.10)$$

where the operator ($\bar{\ }$) is defined as

$$\bar{\beta} = \frac{1}{2\Delta t} \int_{t - \Delta t}^{t + \Delta t} \beta dt. \qquad (4.11)$$

We define CFL time-step limits Δt_g and Δt_m for the external gravity modes and the fastest moving Rossby mode, respectively, and choose a time step Δt such that $\Delta t < \Delta t_m$. Since the gravity modes are assumed to be small in amplitude (quasilinear) and contain only a small fraction of the total energy, the RST vary slowly over the time scale of the Rossby mode. These can be computed once every other Δt time step. Thus the RST are evaluated at time t, and Eqs. (4.8)–(4.10) become

$$u(t + \Delta t) - u(t - \Delta t) + 2\Delta t \delta_x \bar{\phi} = 2\Delta t A_u(t), \qquad (4.12)$$

$$v(t - \Delta t) - v(t - \Delta t) + 2\Delta t \delta_y \bar{\phi} = 2\Delta t A_v(t), \qquad (4.13)$$

$$\phi(t - \Delta t) - \phi(t - \Delta t) + 2\Delta t \Phi \bar{D} = 2\Delta t A_\phi(t). \qquad (4.14)$$

However, the terms on the left-hand side of Eqs. (4.12)–(4.14), namely, the pressure gradient and divergence terms, vary over the time scales determined by all the modes.

The integrals on the left side of (4.12)–(4.14) can be approximated by

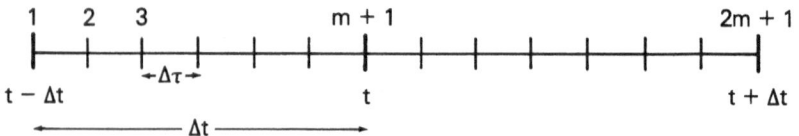

Fig. 4-2 Time intervals Δt divided into subintervals $\Delta\tau$ with $\Delta t = m\Delta\tau$.

using finite differences in time. The most commonly used approximation for (4.11) are

$$\bar{\beta} = \frac{1}{2\Delta t}\int_{t-\Delta t}^{t+\Delta t}\beta dt \doteq \beta(t) \tag{4.15}$$

$$\doteq \tfrac{1}{2}[\beta(t + \Delta t) + \beta(t - \Delta t)], \tag{4.16}$$

where β represents any dependent variable. Equations (4.15) and (4.16) provide a very good approximation for $\bar{\beta}$ if β varies linearly in the interval between $t - \Delta t$ and $t + \Delta t$. If the variation is not linear, then a more accurate approximation can be defined as follows. We subdivide the time interval $2\Delta t$ into m subintervals of length $2\Delta\tau$ (Fig. 4-2). Then we can use approximation (4.15) in each subinterval to get

$$\bar{\beta} = \tilde{\beta} = \frac{1}{m}\sum_{n=1}^{m}\beta^{n}, \tag{4.17}$$

where $m = \Delta t/\Delta\tau$ and

$$\beta^{n} = \beta(t - \Delta t + n\Delta\tau). \tag{4.18}$$

Approximations (4.15) and (4.16) are the basis of the conventional explicit and semi-implicit methods, respectively. Approximation (4.17) gives rise to what we call the split-explicit method. In what follows we will use a tilde instead of a bar for approximation (4.17), so as to distinguish the split-explicit method from the other ones.

We now examine the temporal truncation error and estimate the correction needed. Equations (4.12)–(4.14) can be written as

$$u(t + \Delta t) + 2\Delta t\delta_{x}[\bar{\phi} - \phi(t)] = u^{\mathrm{ex}}(t + \Delta t), \tag{4.19}$$

$$v(t + \Delta t) + 2\Delta t\delta_{y}[\bar{\phi} - \phi(t)] = v^{\mathrm{ex}}(t + \Delta t), \tag{4.20}$$

$$\phi(t + \Delta t) + 2\Delta t\Phi[\bar{D} - D(t)] = \phi^{\mathrm{ex}}(t + \Delta t), \tag{4.21}$$

where the superscript "ex" denotes values computed using explicit time integration over $2\Delta t$. These take the form

$$u^{\mathrm{ex}}(t + \Delta t) = u(t - \Delta t) + 2\Delta t[A_{u}(t) - \delta_{x}\phi(t)], \tag{4.22}$$

$$v^{\mathrm{ex}}(t + \Delta t) = v(t - \Delta t) + 2\Delta t[A_{v}(t) - \delta_{y}\phi(t)], \tag{4.23}$$

$$\phi^{ex}(t + \Delta t) = \phi(t - \Delta t) + 2\Delta t[A_\phi(t) - \Phi D(t)]. \tag{4.24}$$

The second terms on the left of (4.19)–(4.21) represent the deviation of the variable averaged over $2\Delta t$ from the value at the center of the interval. They are proportional to the second derivatives of the variables with respect to t and therefore measure the curvature of the variables in the time interval $2\Delta t$. For time steps less than Δt_g, the variables can be approximated by a straight line between $t - \Delta t$ and $t + \Delta t$; then the correction terms vanish, giving the explicit integration scheme. These correction terms must be added to the predicted values when using the semi-implicit or the split-explicit method.

For the semi-implicit scheme, the correction term for any variable can be written as

$$\bar{\beta} - \beta(t) = \tfrac{1}{2}[\beta(t + \Delta t) + \beta(t - \Delta t)] - \beta(t). \tag{4.25}$$

Now it should be clear that the variables at time t are related implicitly to each other. Since these equations are linear at time $t + \Delta t$, all but one of the variables can be eliminated. From (4.19) and (4.20) we obtain

$$\bar{D} - D(t) + \Delta t(\delta_x^2 + \delta_y^2)[\bar{\phi} - \phi(t)] \\ = \tfrac{1}{2}[D^{ex}(t + \Delta t) + D(t - \Delta t)] - D(t). \tag{4.26}$$

Eliminating $\bar{D} - D(t)$ from (4.21) and (4.26), we obtain

$$\bar{\phi} - \phi(t) - (\Delta t)^2 \Phi(\delta_x^2 + \delta_y^2)[\bar{\phi} - \phi(t)] \\ = \tfrac{1}{2}[\phi^{ex}(t + \Delta t) + \phi(t - \Delta t)] - \phi(t) - \Delta t \Phi\{\tfrac{1}{2}[D^{ex}(t + \Delta t) \tag{4.27} \\ + D(t - \Delta t)] - D(t)\}.$$

The right side of (4.27) involves the known explicit predictions and terms at t and $t - \Delta t$, and therefore can be computed. Equation (4.27) is a Helmholtz equation which can be solved by using any one of the available techniques.

For the split-explicit method, we choose Δt to satisfy the CFL criterion for the gravity modes, $\Delta t < \Delta t_g$. Then the correction terms can be written as

$$\bar{\beta} - \beta(t) = \frac{1}{m} \sum_{n=1}^{m} \beta^n - \beta(t), \tag{4.28}$$

where we must now describe how the subinterval values (4.18) are computed. We rewrite the averaged Eqs. (4.12)–(4.14) using $\Delta\tau$ instead of Δt, and employing the explicit average (4.15) on the left-hand side:

$$u^{n+1} - u^{n-1} + 2\Delta\tau\delta_x\phi^n = 2\Delta\tau A_u(t), \tag{4.29}$$

$$v^{n+1} - v^{n-1} + 2\Delta\tau\delta_y\phi^n = 2\Delta\tau A_v(t), \tag{4.30}$$

$$\phi^{n+1} - \phi^{n-1} + 2\Delta\tau\phi D^n = 2\Delta\tau A_\phi(t). \tag{4.31}$$

We can recover (4.19)–(4.21) by summing these equations over the long time step Δt. Equations (4.29)–(4.31) can be used to compute the correction terms for the split-explicit method. We rewrite (4.29)–(4.31) as

$$[u^{n+1} - u(t)] - [u^{n-1} - u(t)] + 2\Delta\tau\delta_x[\phi^n - \phi(t)]$$
$$= \frac{1}{m}[u^{ex}(t + \Delta t) - u(t - \Delta t)], \tag{4.32}$$

$$[v^{n+1} - v(t)] - [v^{n-1} - v(t)] + 2\Delta\tau\delta_y[\phi^n - \phi(t)]$$
$$= \frac{1}{m}[v^{ex}(t + \Delta t) - v(t - \Delta t)], \tag{4.33}$$

$$[\phi^{n+1} - \phi(t)] - [\phi^{n-1} - \phi(t)] + 2\Delta\tau\Phi[D^n - D(t)]$$
$$= \frac{1}{m}[\phi^{ex}(t + \Delta t) - \phi(t - \Delta t)]. \tag{4.34}$$

From (4.32) and (4.33), the divergence equation is

$$[D^{n+1} - D(t)] - [D^{n-1} - D(t)] + 2\Delta\tau(\delta_x^2 + \delta_y^2)[\phi^n - \phi(t)]$$
$$= \frac{1}{m}[D^{ex}(t + \Delta t) - D(t - \Delta t]. \tag{4.35}$$

Since the RST are known, Eqs. (4.34) and (4.35) can be marched forward over the subintervals to obtain the deviations. For the first subinterval a forward time integration scheme can be used. When these deviations are averaged at the even subtime points we obtain the correction terms defined in (4.19)–(4.21).

Comparison of the three integration schemes: In both the explicit and split-explicit methods for barotropic models, the pressure gradient and the divergence (the terms on the left side) are advanced over the same time interval. In the split-explicit method the RST are computed once every m subtime steps (or once every Δt). Because the linear terms require a small fraction of the time needed for the nonlinear terms, the split-explicit method is nearly m times faster than the explicit method. Since the time steps do not exceed the respective CFL limits for either the meteorological or gravity modes, the results in the two models are nearly equal.

In the semi-implicit model, even though the RST are computed over time steps Δt, some of the computing time saved in using the larger time steps is lost in solving the resulting Helmholtz equation. Since Eq. (4.27) is a second-order difference equation, it can be solved very efficiently by using direct solvers such as the one developed by Madala (1978), which is described in Section 7.2. In this case the semi-implicit method is only slightly inferior to the split-explicit method. This is true only when the staggered system shown in Fig. 4-1 is used to approximate the derivatives. For a nonstaggered system the Helmholtz equation allows more than one decoupled solution, and at present there are no direct methods available to solve it. For a nonstaggered system the equations, which are weakly diagonally dominant, have to be solved by iterative techniques.

The split-explicit method also has significant advantages over the semi-implicit method when a nested grid system such as the one described in Madala and Piacsek (1975) is used, because the Helmholtz equations in the latter cannot be solved directly on a nested grid.

Kreiss and Oliger (1972) have shown that the gravity waves are retarded in using the implicit method more than is required to keep the computation stable. If we use time steps very close to Δt_{max}, the gravity waves will propagate nearly at the phase speed of the Rossby modes. This spurious retardation has an adverse effect on the geostrophic adjustment of the model. The split-explicit method is thus superior to the other two methods.

As mentioned in Section 4.1, the term "split" in the split-explicit technique does not mean the use of separate time steps for different terms as in Gadd's (1978) formulation. Instead, it refers to the separation of the motion in terms of the eigenmodes. For barotropic models, two time steps are used, Δt for the Rossby modes and $\Delta \tau$ for the normal (gravity) modes. In a baroclinic model, separate time steps are employed for each of the gravity modes. This procedure is described in the next section.

4.3 Time Integration Schemes for Baroclinic Models

A second-order-accurate, quadratic conservative finite-difference form of the equations of motion governing a baroclinic model (one in which pressure depends on density, temperature, and any other independent thermodynamic variables which may be present) with normalized pressure as the vertical coordinate is given (Madala, 1981) by

$$\frac{\partial (P_s \mathbf{u})}{\partial t} + \delta_x \Phi = \mathbf{A}_u, \tag{4.36}$$

$$\frac{\partial (P_s \mathbf{v})}{\partial t} + \delta_y \Phi = \mathbf{A}_v, \tag{4.37}$$

$$\frac{\partial (P_s \mathbf{T})}{\partial t} + \mathfrak{M}_2 \cdot \mathbf{D} = \mathbf{A}_T, \tag{4.38}$$

$$\frac{\partial P_s}{\partial t} + \mathbf{N}_1 \cdot \mathbf{D} = 0, \tag{4.39}$$

$$\Phi = \mathfrak{M}_1 \cdot \mathbf{T}, \tag{4.40}$$

and

$$\Phi = P_s[\phi + RT^* - \phi^*], \tag{4.41}$$

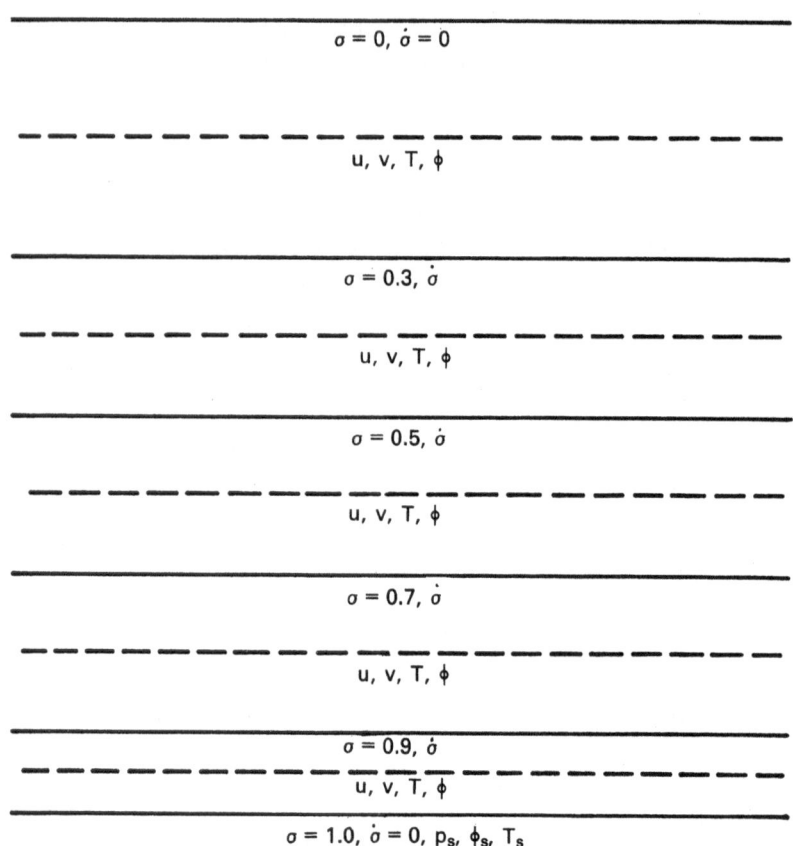

Fig. 4-3 Vertical layering of baroclinic model. Here the subscript s denotes values at the surface of the earth ($\sigma = 1.0$). The nonuniform mesh spaces are drawn to scale, the smallest being the one closest to the surface.

where the ith elements of the vectors $\mathbf{u}$, $\mathbf{v}$, $\mathbf{D}$, $\mathbf{T}$, and ϕ represent the horizontal components of the velocity, mass divergence, temperature, and geopotential, respectively, in the ith layer (Fig. 4-3). The matrices $\mathfrak{M}_1$ and $\mathfrak{M}_2$ and the vector $\mathbf{N}_1$ are independent of time and the x and y coordinates.

The homogeneous solutions of (4.36)–(4.38) contain only the gravity waves. These are the natural gravity waves (eigenmodes) of the numerical model. There are as many natural modes as there are layers in the model.

The vertical structure and phase velocity of the eigenmodes can be obtained by solving the governing equations. Eliminating all the variables except $\boldsymbol{\Phi}$ from the left side of Eqs. (4.36)–(4.40) and neglecting the terms on the right, we obtain

$$\frac{\partial^2 \mathbf{\Phi}}{\partial t^2} - \mathfrak{M}_3 \cdot (\delta_x^2 + \delta_y^2)\mathbf{\Phi} = 0, \tag{4.42}$$

where

$$\mathfrak{M}_3 = \mathfrak{M}_1 \cdot \mathfrak{M}_2 + (\mathbf{RT} - \phi^*)\mathbf{N}_1. \tag{4.43}$$

If $\mathfrak{E}$ represents the eigenvector matrix (with each column representing an eigenvector) of $\mathfrak{M}_3$, then multiplying (4.42) by the inverse of $\mathfrak{E}$ we find

$$\frac{\partial^2 \mathbf{e}}{\partial t^2} - \mathbf{\Lambda} \cdot (\delta_x^2 + \delta_y^2)\mathbf{e} = 0, \tag{4.44}$$

where $\mathbf{e} = \mathfrak{E}^{-1} \cdot \mathbf{\Phi}$. The diagonal matrix $\mathbf{\Lambda}$ is defined as

$$\mathbf{\Lambda} = \mathfrak{E}^{-1} \cdot \mathfrak{M}_3 \cdot \mathfrak{E}; \tag{4.45}$$

its diagonal elements are the eigenvalues of $\mathfrak{M}_3$. If e_i represents the ith element of the vector $\mathbf{e}$, then Eq. (4.44) can also be written as

$$\frac{\partial^2 e_i}{\partial t^2} - \lambda_i(\delta_x^2 + \delta_y^2)e_i = 0, \tag{4.46}$$

where λ_i is the ith diagonal element of $\mathbf{\Lambda}$. Equation (4.46) is a wave equation for the amplitude of the ith mode, whose characteristic phase speed is $\lambda_i^{1/2}$. The vertical structure of this mode is given by the elements of the ith column of $\mathfrak{E}$.

For a five-layer numerical model (Fig. 4-3) employing the mean tropical atmosphere, the five eigenmodes have characteristic phase velocities of approximately 300, 60, 20, 10, and 4 ms^{-1}. The vertical structure of these modes is given in Fig. 4-4. The first mode, the surface mode, has nearly constant amplitude in the vertical direction. The other four modes are the internal gravity modes. The ith internal mode has $i-1$ zeros in the vertical direction.

The RST in (4.36)–(4.38), which are neglected in deriving the eigenmodes of the model, not only modify the amplitude and phase speed of the gravity waves, but also allow Rossby modes as solutions. From the governing equations we can derive the equations for mass divergence $\mathbf{D}$ and the generalized geopotential $\mathbf{\Phi}$. They are

$$\frac{\partial \mathbf{D}}{\partial t} + (\delta_x^2 + \delta_y^2)\mathbf{\Phi} = (\delta_x \mathbf{A}_u + \delta_y \mathbf{A}_v) \tag{4.47}$$

and

$$\frac{\partial \mathbf{\Phi}}{\partial t} + \mathfrak{M}_3 \cdot \mathbf{D} = \mathfrak{M}_1 \cdot \mathbf{A}_T. \tag{4.48}$$

The natural gravity modes described above form a complete set of eigenfunctions satisfying the boundary conditions of the numerical model. There-

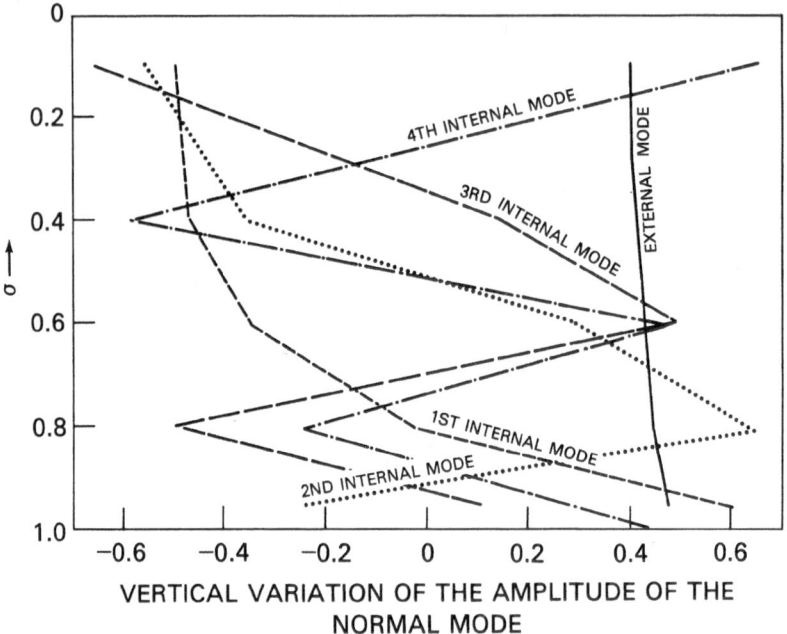

Fig. 4-4 Vertical variation of the five gravity eigenmodes in the five-layer baroclinic model.

fore, the variation of the dependent variables can be expressed as linear combinations of the structure functions (the respective columns of the matrix $\mathfrak{E}$) of these modes. These expansions can be written as

$$\mathbf{u} = \mathfrak{E} \cdot \mathbf{a},$$

$$\mathbf{v} = \mathfrak{E} \cdot \mathbf{b},$$

$$\mathbf{T} = \mathfrak{E} \cdot \mathbf{c},$$

$$\mathbf{D} = \mathfrak{E} \cdot \mathbf{d},$$

$$\mathbf{\Phi} = \mathfrak{E} \cdot \mathbf{e}, \qquad (4.49)$$

where the elements of the coefficient vectors $\mathbf{a}, \mathbf{b}, \mathbf{c}, \mathbf{d}$, and $\mathbf{e}$ are the amplitudes of the eigenmodes. We multiply Eqs. (4.36)–(4.37) and Eq. (4.48) by $\mathfrak{E}^{-1}$ and use (4.49) to obtain the following spectral equations for the wave amplitudes:

$$\frac{\partial(P_s\mathbf{a})}{\partial t} + \delta_x\mathbf{e} = \mathfrak{E}^{-1} \cdot \mathbf{A}_u, \qquad (4.50)$$

$$\frac{\partial(P_s\mathbf{b})}{\partial t} + \delta_y\mathbf{e} = \mathfrak{E}^{-1} \cdot \mathbf{A}_v, \qquad (4.51)$$

$$\frac{\partial \mathbf{e}}{\partial t} + \Delta \cdot \mathbf{d} = \mathfrak{E}^{-1} \cdot \mathfrak{M}_1 \cdot \mathbf{A}_T, \tag{4.52}$$

or for the ith mode,

$$\frac{\partial (P_s a_i)}{\partial t} + \delta_x e_i = (\mathfrak{E}^{-1} \cdot \mathbf{A}_u)_i, \tag{4.53}$$

$$\frac{\partial (P_s b_i)}{\partial t} + \delta_y e_i = (\mathfrak{E}^{-1} \cdot \mathbf{A}_v)_i, \tag{4.54}$$

$$\frac{\partial e_i}{\partial t} + \lambda_i d_i = (\mathfrak{E}^{-1} \cdot \mathfrak{M}_1 \cdot \mathbf{A}_T)_i, \tag{4.55}$$

where the subscript i labels the ith element of the vector and $d_i = \delta_x(P_s a_i) + \delta_y(P_s b_i)$. If the RST of (4.53)–(4.55) are known, these equations can be advanced using suitable integration schemes to obtain the required solution. To compute these RST, however, the surface pressure and the spectral components are needed. These equations can be obtained from (4.38) and (4.39) as

$$\frac{\partial (P_s c_i)}{\partial t} + (\mathfrak{E}^{-1} \cdot \mathfrak{M}_2 \cdot \mathfrak{E} \cdot \mathbf{d})_i = (\mathfrak{E}^{-1} \cdot \mathbf{A}_T)_i \tag{4.56}$$

and

$$\frac{\partial P_s}{\partial t} + (\mathbf{N}_1 \cdot \mathfrak{E} \cdot \mathbf{d}) = 0. \tag{4.57}$$

Equations (4.53)–(4.55) govern the propagation of a gravity mode whose characteristic velocity is $\lambda_i^{1/2}$. They also describe some meteorological modes through the RST. Equations (4.53)–(4.55) are similar to (4.1)–(4.3), with Φ replaced by λ_i and the dependent variables replaced by the wave amplitudes. Therefore, the integration schemes described for the barotropic case in the previous section can be readily applied to Eqs. (4.53)–(4.55). As in Section 4.2, we define the time steps Δt and $\Delta \tau$ such that they do not exceed the CFL criteria respectively for the Rossby modes (Δt_m) and the gravity mode ($\Delta \tau_g$). Since the Rossby modes are not separated, Δt is the same for all the eigenmodes, while $\Delta \tau$ varies with mode. It can vary by two orders of magnitude between the external mode and the lowest internal modes. Even though Δt_g may exceed Δt_m, $\Delta \tau$ is not to exceed Δt. This is required to ensure the accuracy of the solution. As stated in Section 4.2, Δt is chosen to be an integer multiple of $\Delta \tau$.

Following the procedure described in Section 4.2, various integration schemes for Eqs. (4.53)–(4.55) can be written as

$$a_i(t + \Delta t) + 2\Delta t \delta_x[\bar{e}_i - e_i(t)] = a_i^{\text{ex}}(t + \Delta t), \tag{4.58}$$

$$b_i(t + \Delta t) + 2\Delta t \delta_y[\bar{e}_i - e_i(t)] = b_i^{\text{ex}}(t + \Delta t), \tag{4.59}$$

$$e_i(t + \Delta t) + 2\Delta t \lambda_i[\bar{d}_i - d_i(t)] = e_i^{\text{ex}}(t + \Delta t), \tag{4.60}$$

where the superscript "ex" represents the values predicted with an explicit scheme over a time step of $2\Delta t$, and where for any variable β_i, the operator $(^-)$ is defined in analogy with (4.15)–(4.17):

$$\bar{\beta}_i = \frac{1}{2\Delta t}\int_{t-\Delta t}^{t+\Delta t} \beta_i \, dt \doteq \beta_i(t) \qquad \text{(explicit method over } 2\Delta t \text{)} \quad (4.61)$$

$$\doteq \frac{1}{m_i}\sum_{n=1}^{m_i} \beta_i^n \quad \text{(split-explicit method over } 2\Delta\tau_i\text{)} \quad (4.62)$$

$$\doteq \frac{\beta_i^{t-\Delta t} + \beta_i^{t-\Delta t}}{2} \quad \text{(semi-implicit method)}. \qquad (4.63)$$

Here $\Delta\tau_i$ represents the subinterval $\Delta\tau$ for the ith mode, and the integer m_i is the ratio of $\Delta\tau$ to $\Delta\tau_i$.

The correction terms $\bar{e}_i - e_i(t)$ and $\bar{d}_i - d_i(t)$ are obtained for the semi-implicit method by solving the Helmholtz equation (see Section 4.2)

$$\bar{e}_i - e_i(t) - \Delta t^2 \lambda_i(\delta_x^2 + \delta_y^2)[\bar{e}_i - e_i(t)] = F_i \qquad (4.64)$$

and the diagnostic equation

$$\bar{d}_i - d_i(t) + \Delta t(\delta_x^2 + \delta_y^2)[\bar{e}_i - e_i(t)]$$
$$= \tfrac{1}{2}[d_i^{ex}(t + \Delta t) + d_i(t - \Delta t)] - d_i(t). \qquad (4.65)$$

For the split-explicit method, these terms are obtained by solving the divergence and geopotential equations over the subinterval $\Delta\tau_i$ in the form

$$[d_i^{n+1} - d_i(t)] - [d_i^{n-1} - d_i(t)] + 2\Delta\tau_i(\delta_x^2 + \delta_y^2)[e_i^n - e_i(t)]$$
$$= \frac{1}{m_i}[d_i^{ex}(t + \Delta t) - d_i(t - \Delta t)] \qquad (4.66)$$

and

$$[e_i^{n+1} - e_i(t)] - [e_i^{n-1} - e_i(t)] + 2\Delta\tau_i\lambda_i[d_i^n - d_i(t)]$$
$$= \frac{1}{m_i}[e_i^{ex}(t + \Delta t) - e_i(t - \Delta t)]. \qquad (4.67)$$

These equations are integrated over the subtime interval $\Delta\tau_i$ and the solution can be averaged to obtain the correction terms. For the explicit scheme over $2\Delta t$, it should be noted that the correction terms vanish.

The equation corresponding to Eq. (4.57) can be written as

$$P_s(t + \Delta t) + 2\Delta t \mathbf{N}_1 \cdot \mathfrak{E} \cdot [\bar{\mathbf{d}} - \mathbf{d}(t)] = P_s^{ex}(t + \Delta t). \qquad (4.68)$$

Since the gravity modes are separated, during the basic time step Δt the type of the integration scheme and the time step $\Delta\tau_i$ can be varied between the modes. No one scheme may be computationally efficient for all the modes. For the modes for which $\Delta t_m \leq \Delta t_g$, the explicit scheme is stable with time steps given by the Rossby modes. Therefore, for these modes the use of any other scheme is computationally less efficient. For the high-frequency eigenmodes one can use a semi-implicit scheme with time steps Δt or a split-explicit method with time steps $\Delta \tau_i < \Delta t$ for the gravity terms. As explained in Section 4.2, the split-explicit method is more flexible to use and more accurate than the semi-implicit techniques.

Depending on the time step and time integration scheme, we can now derive four different schemes, namely, the explicit, the semi-implicit, the split semi-implicit, and split-explicit schemes. The first three schemes use the same time step for all the modes. For the explicit scheme all the modes are integrated explicitly with time steps determined by the external gravity mode. Since the internal modes travel significantly slower than the external mode, this scheme is very inefficient for the internal modes. In the semi-implicit model (Robert et al., 1972) all the gravity modes are treated implicitly and use time steps determined by the Rossby mode. However, some of the computing efficiency gained through increasing the time step compared to the explicit method is lost in solving a Helmholtz equation for each mode. Since the slow-moving modes are stable for explicit techniques, the additional time expended in solving Helmholtz equations for these modes makes the method computationally inefficient. This is corrected in the split semi-implicit method (Madala, 1975). In this method, for the eigenmodes moving faster than the meteorological modes, a semi-implicit integration is used.

The computational efficiency and accuracy of the methods described above have been examined by using a multilayer limited-area (i.e., not global) baroclinic model developed at the Naval Research Laboratory. The model has five layers (Fig. 4-3) in the vertical direction bounded by constant-σ surfaces. The thickness of these layers varies approximately from 1 to 3 km. In the horizontal direction, a square domain 3000 km on a side is divided into a uniform grid network of cell size 60 km. The physics of the model includes latent heat release in stable and convectively unstable atmospheres. The exchange of momentum and heat between the atmosphere and the earth's underlying surface is parametrized using a generalized similarity theory.

The model has five gravity eigenmodes whose structure and phase speeds are given in Section 4.3. The only modes that travel faster than the Rossby modes are the external and the first internal gravity modes. The Rossby and the second, third, and fourth internal modes are stable for time steps not exceeding 300 s, while the CFL condition restricts the maximum allowable time steps for the external and first internal gravity modes to 70 and 120 s, respectively. The model is integrated with 300-s time steps using the semi-implicit and split semi-implicit techniques. In the latter method only the first two gravity modes are treated implicitly. The explicit model is integrated using 50-s time steps. The split-explicit method uses 50-s time steps for the external mode, 100-s time steps for the first internal mode, and 300-s time steps for the other three modes.

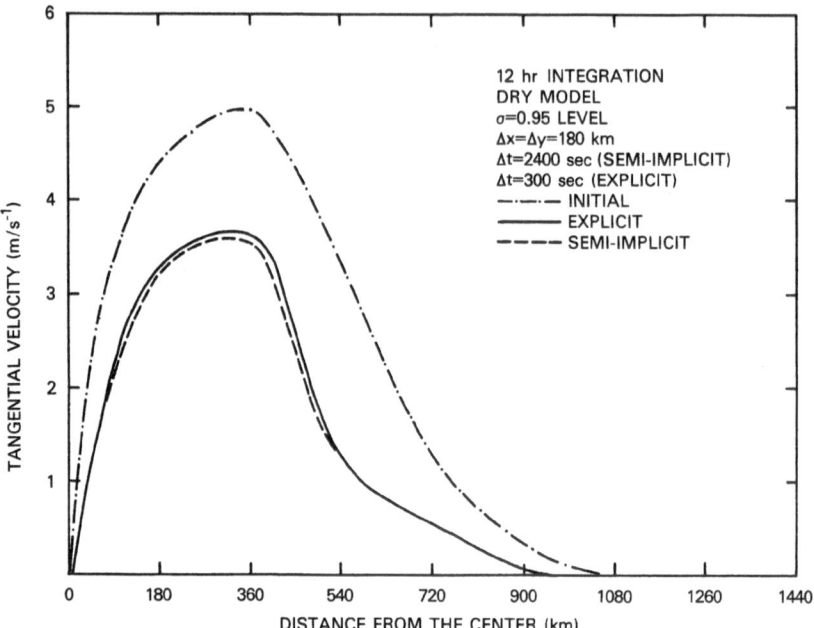

Fig. 4-5 Tangential velocity after 12 h of integration forecast by the tropical cyclone model. Explicit and semi-implicit time differences show nearly identical dissipation of the initial vortex due to skin friction.

The initial conditions for the integration of the model consist of a weak vortex in gradient wind balance (a balance between centrifugal, Coriolis and pressure gradient forces). A dry version of the model is integrated for 12 hr using all four methods. During this period, the initial vortex intensifies slowly with time into a tropical cyclone.

Figures 4-5 and 4-6 describe the time variation of the tangential velocity and the radial velocity at the $\sigma = 0.95$ level from the semi-implicit (dashed curve) and explicit (solid curve) models. Results from the split-explicit integration are not shown here, as they are indistinguishable from the ones obtained by explicit integration. It is clear from these figures that the accuracy of the solution does not suffer by the use of the split-explicit method. The results from the split semi-implicit methods are not shown here. However, they also deviated very little from those obtained by the explicit method. The resulting Helmholtz equations in the semi-implicit and split semi-implicit methods are solved by the SEVP method (Section 7.2) on the TI-ASC computer. The explicit, semi-implicit, split semi-implicit, and split-explicit methods took 160, 48, 40, and 30 min of CPU time, respectively, for 36 hr of integration.

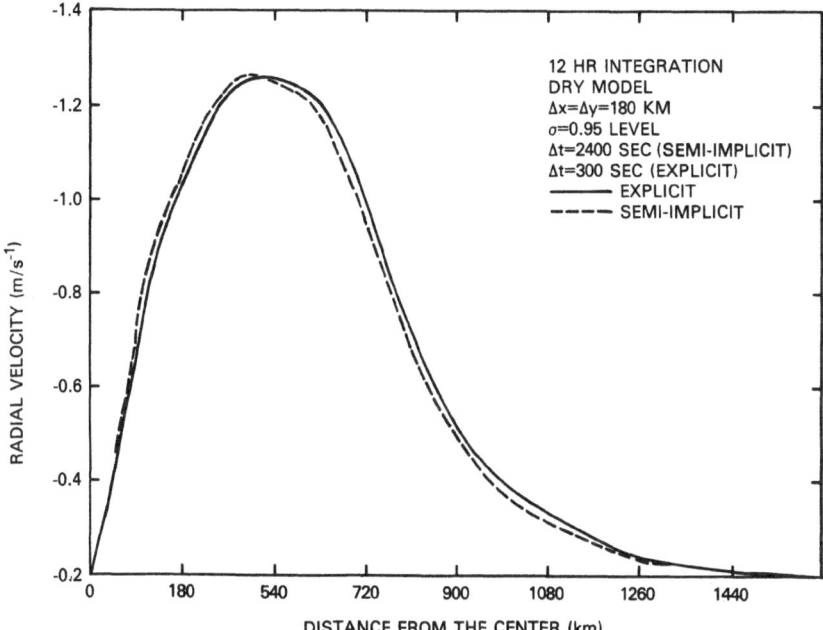

Fig. 4-6 Radial velocities, calculated as in Fig. 4-5. Initial condition was taken to be $v = 0$ (balanced vortex).

4.4 Extension to Ocean Models

The methods described in Section 4.3 are more efficient for ocean models than atmospheric models since the external gravity mode travels about two orders of magnitude faster than the next fastest mode (the first internal mode) allowed in this model, compared with a factor of 10 in atmospheric models. For large-scale ocean circulations the phase speeds of ocean circulations and the internal modes are of the same order of magnitude. Therefore, just by treating the external mode implicitly one can increase the time step by a factor of about 100. The terms governing the external gravity waves are easily separable from those that govern the remaining modes.

The equations governing a Boussinesq hydrostatic baroclinic ocean model can be written as

$$\frac{\partial u}{\partial t} = -\frac{1}{\rho_0}\frac{\partial p}{\partial x} + f_u, \tag{4.69}$$

$$\frac{\partial v}{\partial t} = -\frac{1}{\rho_0}\frac{\partial p}{\partial y} + f_v, \tag{4.70}$$

$$\frac{\partial h}{\partial t} = -H\left\{\frac{\partial [u]}{\partial x} + \frac{\partial [v]}{\partial y}\right\} - \left\{\frac{\partial [u_1 h]}{\partial x} + \frac{\partial [v_1 h]}{\partial y}\right\}, \tag{4.71}$$

$$\frac{\partial p}{\partial z} = -\rho g, \tag{4.72}$$

where u and v are the horizontal components of the velocity, p is total pressure, ρ is the density, ρ_0 is the mean density (assumed to be a function of depth only), and h is the deviation of the ocean depth from its mean value H. The Coriolis terms, nonlinear terms, sources, and sinks of momentum are represented by f_u and f_v. The square brackets denote vertical averages over the mean depth of the ocean, i.e., for an arbitrary fluid variable,

$$[\beta] = \frac{1}{H} \int_0^H \beta(z)dz. \tag{4.73}$$

The equation of continuity, equation for salinity, the thermodynamic equations, and the equation of state are omitted here since these are not required for separation of modes. The total pressure can be decomposed into three components, namely,

$$p(z) = p_a + \rho_0^{(1)} gh + p' + \rho_0 gz, \tag{4.74}$$

where the terms on the right represent respectively the atmospheric pressure at the surface, the hydrostatic pressure due to changes in the height of the free surface, the pressure due to baroclinicity of the ocean, and that due to the variation with depth; here $\rho_0^{(1)}$ is the mean density in the topmost layer of the model.

Substituting (4.74) into Eqs. (4.69) and (4.70), we obtain

$$\frac{\partial u}{\partial t} = -\frac{1}{\rho_0}\frac{\partial p_a}{\partial x} - g\frac{\partial h}{\partial x} - \frac{1}{\rho_0}\frac{\partial p'}{\partial x} + f_u \tag{4.75}$$

and

$$\frac{\partial v}{\partial t} = -\frac{1}{\rho_0}\frac{\partial p_a}{\partial y} - g\frac{\partial h}{\partial y} - \frac{1}{\rho_0}\frac{\partial p'}{\partial y} + f_v. \tag{4.76}$$

The external gravity modes are primarily governed by the first two terms on the right of Eqs. (4.75) and (4.76) and the first term on the right side of Eq. (4.71). The other terms in these equations vary on the time scales given by the internal gravity modes and the large-scale circulations. Averaging Eqs. (4.75) and (4.76) over the mean depth of the ocean, we obtain

$$\frac{\partial [u]}{\partial t} = -\frac{1}{\rho_0}\frac{\partial p_a}{\partial x} - g\frac{\partial h}{\partial x} - \frac{1}{\rho_0}\frac{\partial [p']}{\partial x} + [f_u] \tag{4.77}$$

and

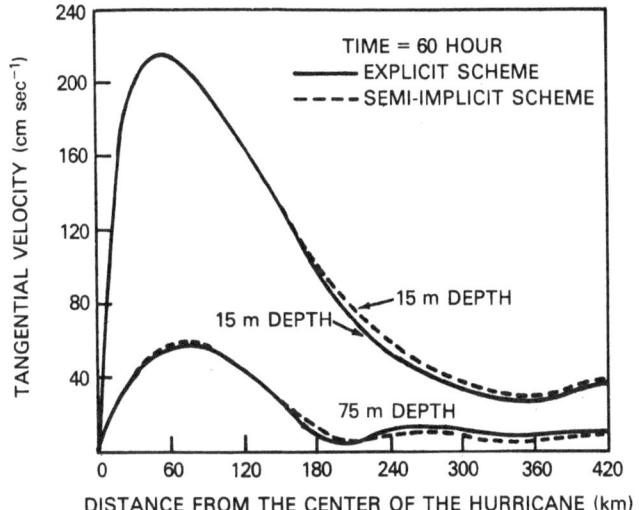

Fig. 4-7 A comparison of the tangential velocities (at 15 and 75 m depths) obtained at the end of 60 h of integration of the semi-implicit and explicit 10-layer ocean models.

$$\frac{\partial [v]}{\partial t} = -\frac{1}{\rho_0}\frac{\partial p_a}{\partial y} - g\frac{\partial h}{\partial y} - \frac{1}{\rho_0}\frac{\partial [p']}{\partial y} + [f_v].\tag{4.78}$$

These two equations together with Eq. (4.71) determine the external gravity wave. For the split-explicit method these equations can be solved for $[u]$, $[v]$, and $[h]$ using time steps given by the CFL condition for the external gravity mode. For an implicit treatment, the mean divergence $[D]$ is eliminated from the equations, yielding a Helmholtz equation for h.

If u' and v' represent the deviations of u and v from $[u]$ and $[v]$, the equations governing them can be obtained by subtracting Eqs. (4.77) and (4.78) from Eqs. (4.75) and (4.76), respectively. The resulting equations are

$$\frac{\partial u'}{\partial t} = -\frac{1}{\rho_0}\frac{\partial p'}{\partial x} - \frac{1}{\rho_0}\frac{\partial [p']}{\partial x} + f_u - [f_u]\tag{4.79}$$

and

$$\frac{\partial v'}{\partial t} = -\frac{1}{\rho_0}\frac{\partial p'}{\partial y} + \frac{1}{\rho_0}\frac{\partial [p']}{\partial y} + f_v - [f_v].\tag{4.80}$$

Since these contain only those terms which govern the slow-moving modes, they can be integrated explicitly using time steps about 100 times longer than those given by the CFL condition for the external mode.

A 10-layer ocean model treating the external mode implicitly and all the

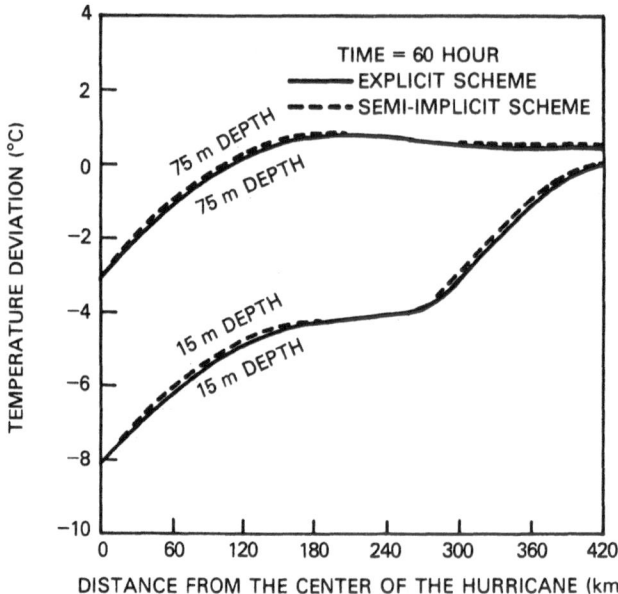

Fig. 4-8 A comparison of the deviatory temperatures (at 15 and 75 m depths) obtained at the end of 60 h of integration of the semi-implicit and explicit 10-layer ocean models.

other modes explicitly has been described by Madala and Piacsek (1977). Figures 4-7 and 4-8 illustrate the comparison of the semi-implicit and explicit time integrations for the azimuthal velocity and temperature deviations, respectively, at various depths. The results are illustrated as a function of radial distance at 60 hr of elapsed real time. The velocity results show appreciable deviation only near the boundaries, particularly in the lower layer. The errors for h in the upper layer amount to less than 10%, but amount to about 25% in the lower layer. The temperature results show a smaller error, a few percent, in all regions of the flow. The velocity discrepancies are attributed to the different arrival times and reflection of the surface waves as treated by the semi-implicit and explicit model, respectively. Sponge layers containing a Rayleigh viscosity can be inserted to reduce reflection effects. The errors seem to have the same absolute size in each layer and can be attributed to surface effects manifested in layers 3 and 7. The amount of computer time saved using the present method is a factor of 12; the time-step difference, a factor of 15, was slightly offset by the time necessary to solve the single two-dimensional Helmholtz equation for the free surface height.

CHAPTER 5

A One-Dimensional Lagrangian Code for Nearly Incompressible Flow

5.1 Difficulties Encountered in Lagrangian Methods

Lagrangian methods offer the most natural approach to transient hydrodynamics problems which contain free surfaces, interfaces, sharp gradients, or boundaries. The advantage of the Lagrangian formulation lies in the fact that fluid elements are advected with the flow. The grid points which define surfaces remain at those surfaces, and grid points near the surface are advected with velocities nearly that of the surface. The surface can therefore remain optimally resolved over its entire extent for long times, permitting maximum accuracy in the formulation of boundary conditions and minimizing the numerical diffusion across the surface. More importantly, the nonlinear convective terms are not present in the Lagrangian formulation, resulting in higher accuracy and less stringent resolution requirements.

However, the greatest strength of the Lagrangian approach is also its greatest weakness. The advection of the mesh with the flow leads to large mesh deformations and a corresponding decrease in accuracy. To illustrate, we will use the one-dimensional mesh shown in Fig. 5-1, which contains a surface located at the ith grid point. We can expand in both the forward and backward directions about the point i to obtain a Taylor series for the fluid variable f:

$$f_{i+1} = f_i + \frac{\partial f}{\partial x_i}\Delta x + \frac{1}{2}\frac{\partial^2 f}{\partial x_i^2}\Delta x^2 + \frac{1}{6}\frac{\partial^3 f}{\partial x^3}\Delta x^3 + O(\Delta x^4) \qquad (5.1)$$

and

$$f_{i-1} = f_i - \frac{\partial f}{\partial x_i}\Delta x' + \frac{1}{2}\frac{\partial^2 f}{\partial x_i^2}\Delta x'^2 - \frac{1}{6}\frac{\partial^3 f}{\partial x^3}\Delta x'^3 + O(\Delta x'^4), \qquad (5.2)$$

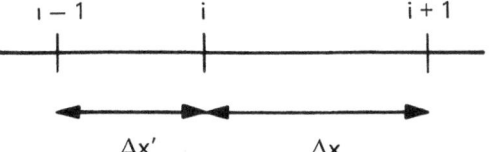

Fig. 5-1 One-dimensional nonuniform mesh. The ith grid point represents the location of a surface.

where $\Delta x = x_{i+1} - x_i$ and $\Delta x' = x_i - x_{i-1}$. Of course, the forward and backward difference approximations can be obtained directly from Eqs. (5.1) and (5.2):

$$\left(\frac{\partial f}{\partial x_i}\right)^+ = \frac{f_{i+1} - f_i}{\Delta x} + O(\Delta x) \tag{5.3}$$

and

$$\left(\frac{\partial f}{\partial x_i}\right)^- = \frac{f_i - f_{i-1}}{\Delta x'} + O(\Delta x'). \tag{5.4}$$

Both are only first-order accurate. If $\Delta x' = \Delta x$ and Eq. (5.2) is subtracted from (5.1), the result is

$$f_{i+1} - f_{i-1} = 2\frac{\partial f}{\partial x_i}\Delta x + \frac{1}{3}\frac{\partial^3 f}{\partial x_i^3}\Delta x^3 + O(\Delta x^5). \tag{5.5}$$

The centered difference approximation is then

$$\frac{\partial f}{\partial x_i} = \frac{f_{i+1} - f_{i-1}}{2\Delta x} + O(\Delta x^2), \tag{5.6}$$

which is second-order accurate. For $\Delta x' \neq \Delta x$, Eq. (5.5) becomes

$$f_{i+1} - f_{i-1} = \frac{\partial f}{\partial x_i}(\Delta x + \Delta x') + \frac{1}{2}\frac{\partial^2 f}{\partial x_i^2}(\Delta x^2 - \Delta x'^2) + O(\Delta x^3). \tag{5.7}$$

Therefore,

$$\frac{\partial f}{\partial x_i} = \frac{f_{i+1} - f_{i-1}}{(\Delta x + \Delta x')} - \frac{1}{2}\frac{\partial^2 f}{\partial x^2}(\Delta x - \Delta x') + O[\max(\Delta x^2, \Delta x'^2)], \tag{5.8}$$

which now is no longer explicitly second-order accurate. We can define the average grid spacing $\overline{\Delta x} = \frac{1}{2}(\Delta x + \Delta x')$, as shown in Fig. 5-2. Then

$$\Delta x = \overline{\Delta x} + \delta x, \tag{5.9}$$
$$\Delta x' = \overline{\Delta x} - \delta x. \tag{5.10}$$

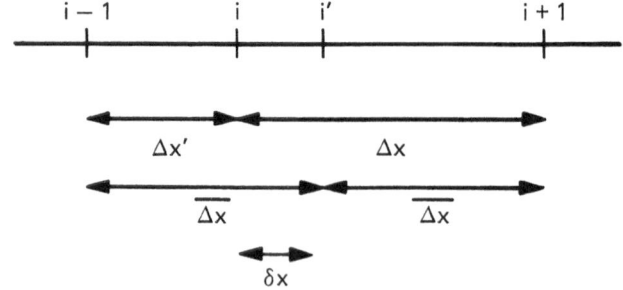

Fig. 5-2 Average grid spacing $\overline{\Delta x}$ for the mesh segment shown in Fig. 5-1. The point $x_{i'}$ is located at the center of the interval (x_{i-1}, x_{i+1}).

After substitution, Eq. (5.8) becomes

$$\frac{\partial f}{\partial x_i} = \frac{f_{i+1} - f_{i-1}}{2\overline{\Delta x}} - \frac{1}{2}\frac{\partial^2 f}{\partial x^2}\frac{(4\overline{\Delta x}\delta x)}{2\overline{\Delta x}} + O[\max(\Delta x^2, \Delta x'^2)] \qquad (5.11)$$

or

$$\frac{\partial f}{\partial x_i} = \frac{f_{i+1} - f_{i-1}}{2\overline{\Delta x}} + O[\max(\delta x, \Delta x^2, \Delta x'^2)]. \qquad (5.12)$$

Equation (5.12) is the variable mesh counterpart of Eq. (5.6) and shows that second-order accuracy is retained only if the δx errors are smaller than those associated with Δx^2 and $\Delta x'^2$. That is, the approximation will only have full first-order accuracy for the degenerate case of two superimposed grid points, i.e., $2\delta x = \Delta x$, $\Delta x' = 0$.

The loss in accuracy shown in Eq. (5.12) should be viewed in perspective. Eq. (5.11) applies also to a change in mesh size in an Eulerian grid, so that an attempt at finer resolution near a boundary to achieve greater accuracy for the convective term can lead to a worsening of error. The advection of the grid in the Lagrangian case generally results only in a gradual loss in accuracy, as implied by Eq. (5.11). Therefore, a point which migrates too near a boundary may simply be deleted. At the time of the point deletion, diffusion occurs back toward the interior, not across the boundary. In the Eulerian case, on the other hand, the more finely resolved part of the grid must follow the surface, so that many more additions or deletions of points would be necessary. Each time the mesh is rezoned the points have to be relabeled and the calculation reinitialized so that the regridded problem can be integrated in a manner which vectorizes.

Another kind of problem encountered in carrying out both Lagrangian and Eulerian fluid calculations arises when the physical phenomena of interest vary slowly compared to the time t_s required by sound waves to cross the system. This regime is often termed "nearly incompressible" because $\nabla \cdot \mathbf{v}$ is small compared with t_s^{-1}. Nevertheless, substantial compressions and expansions can occur since their effects accumulate over many sound transit times. The simplest example of this type of problem is the dynamics of a gas bubble rising through water. The whistle problem mentioned in Section 2.5 represents a different kind of "nearly incompressible" flow in which it is impossible to ignore compressibility. Other examples are provided by imploding pellets and liners. Here, high-density fluids such as compressed DT, heavy pusher shells, and liquid metal rotating liners converge in a way which is often shock-free, yet requires the use of a nontrivial equation of state.

Particularly in implosion hydrodynamics problems, it is often required that sharp interfaces be maintained between distinct materials or between materials in widely disparate states. In Eulerian computations the numerical diffusion needed for stability (cf. Section 3.1) spreads such an interface over several zones and renders knowledge of its location uncertain. We are thus driven to seek a Lagrangian technique for solving the fluid equations, one which is not subject to the Courant time-step restriction,

$$\delta t < \frac{\delta x}{|v| + c}, \tag{5.13}$$

associated with the characteristics $C_\pm$ [Eq. (2.52)], but instead satisfies the less stringent condition

$$\delta t < \frac{\delta x}{|v|} \tag{5.14}$$

associated with C_0 [Eq. (2.34)]. The motivation here—a desire to relax the stringent time-step limitation introduced by a small number of rapidly propagating modes—is very similar to that which gave rise in the preceding chapter to modification of conventional explicit Eulerian techniques applied to models of the upper ocean and the atmosphere.

Conventional Lagrangian codes (e.g., MAGPIE of Steinberg, Kidder, and Cecil, 1966) have proven very accurate in supersonic flow calculations (shocks) involving distinct phases separated by sharp boundaries. A general purpose one-dimensional Lagrangian code should embody two additional features, aimed at alleviating the two types of difficulty just described: a fully Lagrangian adaptive rezone package to maintain resolution in regions of sharp gradients or where zone sizes become unacceptably large; and an implicit finite-difference algorithm to circumvent the Courant condition (5.13). We discuss these in turn.

5.2 Adaptive Gridding in a Lagrangian Calculation

The methodology of adaptive gridding in the Eulerian and the Lagrangian representations is intrinsically different. In Eulerian calculations, a general continuous sliding rezone method can be used. ETBFCT (Section 3.2) performs just this sort of regridding. In contrast, Lagrangian calculations require discontinuous injection and removal of cell interfaces to localize the numerical errors and thus retain the full nondiffusive character of the Lagrangian formulation.

Many Lagrangian models resort to an Eulerian rezone to smooth out the computational mesh. As an example of an Eulerian rezone, consider the reactive shock model of Oran and Boris (1979). A general adaptive gridding procedure was developed which utilizes a sliding rezone to automatically follow a shock front where enhanced resolution is required (Fig. 5-3). The region immediately around the shock front and for predetermined distances on each side is gridded with fine, evenly spaced cells. In principle there can be any number of finely spaced regions. The fine spacing transitions smoothly into

CONTINUOUS EULERIAN REZONE

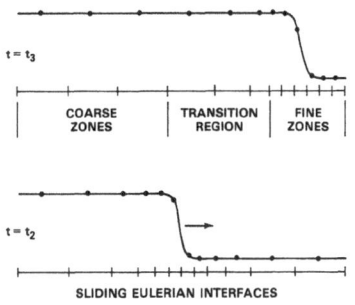

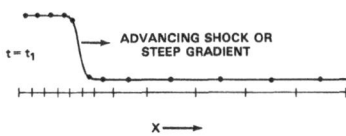

Fig. 5-3 Use of sliding-zone regridding procedure to follow a shock (cf. Figs. 3-13 and 3-14). Fine cells are used in the region of the shock or concentration gradient, while coarse cells are used in the regions of uniform and near-uniform flow.

the more coarsely resolved region to retain the accuracy specified by Eq. (5.12). This smooth variation of cell size is required to maintain a reasonable control over the accuracy of the difference scheme used, as discussed above.

Even when special techniques can be developed to handle rapidly varying cell size (cf. Section 5.3), physical situations arise which sometimes limit the rate at which the cell size may vary. A shock propagating from a region of coarse resolution to one of fine resolution will exhibit nonphysical oscillations if the zone size transition occurs faster than the shock can steepen physically. Propagating across the coarsely gridded region, the shock has a thickness of one cell. In entering the finely gridded region this becomes a ramp several cells across. This ramp steepens up to a shock while shedding oscillations (Boris and Book, 1973). These, while physically correct, are not properly a part of the problem originally posed. Only by transitioning the zone size more slowly than the shock steepens naturally can these "physical unphysical" oscillations be suppressed.

In the Eulerian rezone a prescription must be given for smoothly adjusting the mesh in a global or semiglobal manner. As the shock moves along the length of the system, the finely spaced region is programmed to move with it and may reflect off a boundary wall. The shock front is located by looking for the maximum of $|\Delta p/\Delta r|/\bar{p}$, where p is the pressure, r is a generalized position coordinate, and $\bar{p}$ is an average mass density.

The Eulerian sliding rezone and other global adaptive rezoning procedures run into trouble at interfaces. Resolution is increased where needed by sliding in extra zones from nearby regions where extra resolution is not required.

DISCONTINUOUS LAGRANGIAN REZONE

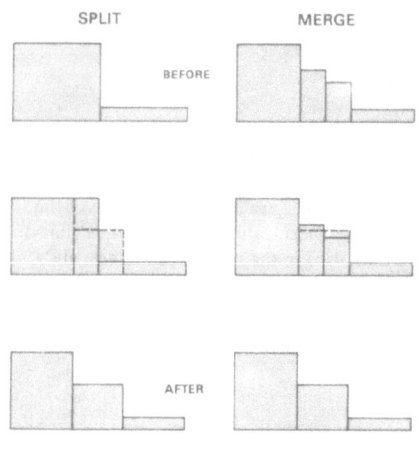

Fig. 5-4 Addition or subtraction of cell boundaries to enhance or diminish resolution.

However, as a shock approaches a liquid–solid interface, cells from the solid cannot slide into the liquid to help without entailing complicated logic to keep the solid and liquid from interpenetrating diffusively.

In a Lagrangian fluid calculation, adaptive gridding is still necessary, but the procedure is more complicated than in the Eulerian case. A fundamental problem arises from the need to arbitrarily concentrate grid points without losing the nondiffusive advantages of the Lagrangian representation. In Eulerian calculations continuous regridding leads to algorithms which are numerically diffusive. While the Lagrangian framework itself provides some natural concentration of grid points in regions where the fluid is compressing, there are situations, such as a flame, where fine resolution is required, but the fluid is expanding. In a flame front calculation, the region requiring the fine resolution actually propagates through the fluid rather than with it. In Eulerian calculations, the grid always moves to some extent through the fluid.

In order to avoid the problems of a diffusive grid in Lagrangian calculations, we have devised a method in which existing computational cells can be split or merged (Fig. 5-4). The addition or subtraction of one or more cell boundaries can be effected without causing numerical diffusion at any of the already existing boundaries. Criteria are programmed so that this process is done automatically in a way which adapts to the resolution needs of the evolving solution. The bookkeeping for this method is complicated since the number of grid points changes and the location in computer memory of data referring to a given physical point in space also changes as cells are subdivided or removed. This method is currently being used in transient flame simulations and in calculations of laser ablation layers where resolution changes over several orders of magnitude are needed.

In the Lagrangian rezone, six conditions on the changing grid and the evolving solution are enforced.

(1) No cell smaller than a minimum size is permitted.
(2) No cell larger than a maximum size is permitted.
(3) Variation of adjacent cell sizes is limited to about 3 : 1.
(4) No relative change in important variables (e.g., temperature and species densities) from one cell to the next is permitted to exceed a maximum percentage.
(5) No relative change in important variables less than a minimum percentage is tracked.
(6) No rezoning across a material interface is permitted.

These conditions interact and even conflict in a complicated way, so a hierarchy of careful tests must be performed to determine where rezoning is needed and permitted. The rezoning is accomplished by injecting new cell boundaries into the interior of existing cells and by removing unneccessary boundaries between two existing cells. The number of cells change during these "split" and "merge" operations (so reordering of data in the computer memory is recommended), but only local restructuring occurs this way. To maintain the Lagrangian nature of the calculation, material and energy are not made to flow across cell boundaries, as in global rezone procedures.

In our procedure the governing physical variable gradients generally determine whether better resolution is required in the vicinity of each cell boundary or whether larger cells would be permissible. Any cell split or merge operation suggested by the combined gradient criteria (conditions 4 and 5) is tested before implementation to determine if the cell size criteria (conditions 1 and 2) are violated. When the required resolution change would cause violation of the cell size variation limit (condition 3), intermediate-size cells are injected to provide a smoother transition. Of course, all of the criteria may vary with position, material, and time. These one-dimensional operations are, in fact, the analog of the two-dimensional grid restructing algorithms discussed in Section 6.3.

In addition to this simple Lagrangian adaptive rezone, more exotic gridding procedures can be considered for the space–time domain. For example, the time step δt can be made to vary with position. In most finite-difference and finite-element calculations where a time evolution equation is being solved, δt does not depend on position. Each grid point is integrated in locked step and the concept of a time level is well defined. However, the time step is usually determined by stability or accuracy conditions at one point on the grid and is smaller than necessary everywhere else. If each cell could be integrated using a different time step, the total number of point steps for the calculation could be reduced. Thus in an advancing shock calculation, small space and time cells could be clustered only along the advancing shock front.

Although having a δt that depends on location reduces the number of point steps required, a serious price may be exacted by the computational complexity introduced. There is another, perhaps simpler, adaptive gridding option which reduces the total number of point steps needed for the calculation even further than the variable δt–δx approach described above. This approach depends intrinsically on the existence of two asymptotically separable space scales and is called "intermittent embedding" of an adaptive grid. Thus, a finely gridded region is embedded in the macroscopic calculation at intermittent time steps. This is done often enough to update the properties of the finely spaced region. These properties (jump conditions, flame speeds, etc.) are then used as interior boundary conditions for the coarsely spaced macroscopic region.

Since the region which requires resolution is only resolved and followed during a small fraction of the time, a method of connecting and disconnecting this portion of the grid must then be developed. Furthermore, when the fine grid for the flame front is disconnected, a method for retaining the discontinuity in the coarsely resolved calculation has to be found. This latter requirement is clearly problem-dependent (consistent with condition 6 above). Thus, for example, a shock would have different jump and conservation conditions from a flame front.

5.3 The Algorithm and Structure of ADINC

The second desirable modification of Lagrangian methods noted in Section 5.1, an implicit convective-equation solver permitting time steps to exceed the Courant condition, can also be incorporated. Recently a method of doing this has been developed by Boris (1979), by means of an algorithm called ADINC (ADiabatic and INCompressible flows). In this and the following section we describe the algorithm and present the results of a series of test problems illustrating its capabilities.

ADINC solves Eq. (2.1) and (2.3) for mass and momentum transport in one dimension in the form

$$\frac{d\rho}{dt} = -\rho \nabla \cdot \mathbf{v}, \tag{5.15}$$

$$\rho \frac{d\mathbf{v}}{dt} = -\nabla p. \tag{5.16}$$

The energy evolution equation is eliminated by using an adiabatic equation of state in which the entropy $\ln s$ is assumed constant throughout the numerical integration. Nonadiabatic processes such as external heating, thermal conduction, and chemical energy release can be added to Eqs. (5.15) and (5.16) using time step splitting, provided sufficiently short time steps are used to make the splitting procedure accurate. In the reference version of ADINC, reproduced as Appendix C, the equation of state of the fluid in each cell of the calculation is

$$\rho(p,s,\dots) = \rho_c + \left(\frac{p}{s}\right)^{1/\gamma}. \tag{5.17}$$

When $\rho_c = 0$, Eq. (5.17) describes adiabatic compression and expansion of an ideal gas. When $\rho_c \neq 0$, Eq. (5.17) gives an adequate representation of a mildly compressible liquid. Water, for example, has $\rho_c = 1\,\mathrm{g\,cm^{-3}}$ and $s \approx 2.5 \times 10^{11}$ in cgs units, and $\gamma \approx 7$. Thus, in this crude model a pressure of 250 kbars ($2.5 \times 10^{11}\,\mathrm{dyn/cm^2}$) causes a relative compression of order unity.

During an ADINC time step ρ_c, γ, and s are treated as constants; only the variation of ρ with p is considered. ADINC does not use the temperature T anywhere. Equation (5.17) is employed in the form specified. This equation-of-state density is compared to the density derived from the fluid dynamics via (5.15). The difference is iterated to zero using a quadratically convergent implicit solution of (5.16) which delivers an improved pressure approximation. During this iteration the analytic derivative $d\Lambda/dp$ is used, where Λ is the volume of a computational cell. Thus

$$\frac{1}{\Lambda}\frac{d\Lambda}{dp} = -\frac{1}{\gamma\rho p}\left(\frac{p}{s}\right)^{1/\gamma} \tag{5.18}$$

for the particular equation of state (5.17).

The ADINC package is sufficiently modular that replacing (5.17) with another equation of state should be straightforward. Thermochemistry and thermophysical real-gas properties can be included so the effective gas constant γ displays the correct variation with T. More involved equations of state for solid and liquid materials can also be included.

Because finite errors in pressure and density are expected during the iteration process, ADINC uses the equation of state in the form $\rho(p,s,\dots)$. Rather than finding the pressure as a function of ρ, ADINC calculates the fluid density from an approximation to the pressure. For liquids and solids the density is a weak function of the pressure. In the other form $p(\rho,s,\dots)$, errors in density ρ would appear as wild pressure fluctuations. For gases and plasma the two forms are basically of the same accuracy. There is a second related reason for using the equation of state in the form $\rho(p,s,\dots)$. The ADINC package is especially designed to deal with discontinuities in zone size and density. When a gas–solid interface is encountered, the pressure is continuous, but the density need not be. Therefore, finite differences in the pressure are bound to be more accurate than if they are calculated from differences in the density according to $\Delta p = (\partial p/\partial\rho)\Delta\rho$.

Equations (5.15)–(5.18) are solved in the form shown, without nondimensionalization or scaling. The package is designed with cgs units in mind, although these appear nowhere explicitly in the calculation. Nondimensional calculations are therefore possible without modifying the program.

To complete the basic equations, the geometry of the calculation must be

**ADINC GRID STRUCTURE
AND VARIABLE PLACEMENT**

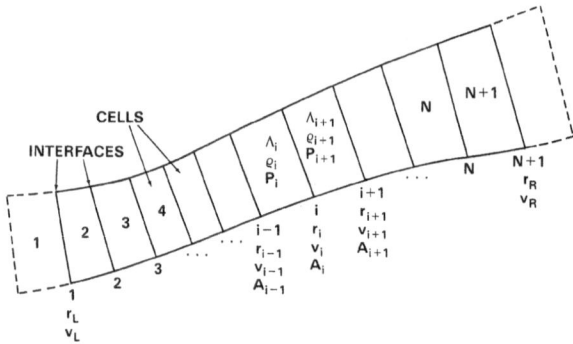

Fig. 5-5 Grid structure and variable definition for the ADINC package. Position r_i and velocity v_i are defined at cell interfaces while density ρ_i and pressure p_i are defined at cell centers. Interface area A_i and cell volume Λ_i are derived from the instantaneous interface positions $\{r_i\}$.

specified. Figure 5-5 shows a general one-dimensional region of variable cross-sectional area $A(r)$. The volume of the region is $V(r) = \int_{r_1}^{r} A(r')dr'$. The numerical algorithm uses interface areas and cell volumes exclusively to convey geometry information. Therefore, substitution of one-dimensional geometries other than those provided for should be straightforward. ADNIC, as presented, allows four one-dimensional systems to be treated by selection of an integer switch α. All four coordinate systems (Cartesian, cylindrical, spherical, nozzle) are special cases of

$$A(r) = G_1 + G_2 r + G_3 r^2 + G_4 r^3 + G_5 r^4, \tag{5.19}$$

$$V(r) = G_1 r + G_2 \frac{r^2}{2} + G_3 \frac{r^3}{3} + G_4 \frac{r^4}{4} + G_5 \frac{r^5}{5}. \tag{5.20}$$

In principle, the volume is ambiguous up to an additive constant, the volume $V(r_1)$. In practice, ADINC deals only with volume differences to determine the incremental volume Λ of a computational cell, so this constant conveniently cancels out.

Figure 5-5 shows a schematic diagram of the computational region treated by the ADINC package. There are N cells of volume Λ_i ($i = 2,3,\ldots,N + 1$) bounded by $N + 1$ interfaces of area A_i($i = 1,\ldots,N + 1$). The interfaces are located at r_i ($i = 1,\ldots,N + 1$), so $A_i = A(r_i)$, where $A(r)$ is given by one of the choices $\alpha = 1, 2, 3, 4$ in Eq. (5.19). The cell volumes $\Lambda_i = V(r_i) - V(r_{i-1})$ are the difference of the volume integral from Eq. (5.20) at the two cell interfaces. In the development to follow we will also need the cell center location

$$R_i = \frac{A_i r_{i-1} + A_{i-1} r_i}{A_i + A_{i-1}}, \tag{5.21}$$

which always lies between r_{i-1} and r_i if the interface areas are positive.

In ADINC the first physical cell is labeled $i = 2$ and lies between interfaces r_1 and r_2. The last physical cell has label $i = N + 1$ and lies between interfaces r_N and r_{N+1}. Cells 1 and $N + 2$ are not used currently by ADINC, but have been left available for future use in complicated boundary conditions or extrapolations. Interface 1 is the left-hand boundary of the system ($r_L = r_1$), and interface $N + 1$ is the right-hand boundary of the system ($r_R = r_{N+1}$). The interfaces are treated in a fully Lagrangian manner; hence the interface velocities $\{v_i\}$ are defined as well as the interface positions $\{r_i\}$. The left-hand and right-hand boundary velocities ($v_L = v_1$ and $v_R = v_1$) are established externally. ADINC integrates the interior interface locations and velocities from one discrete time t to the next, $t + \delta t$, given the masses, entropies, and other conserved cell quantities.

The cell interface positions $\{r_i\}$ satisfy

$$\frac{dr_i}{dt} = v_i, \tag{5.22}$$

which has a straightforward discretization

$$r_i^n = r_i^o + \delta t[\varepsilon_r v_i^o + (1 - \varepsilon_r)v_i^n]. \tag{5.23}$$

In Eq. (5.23), just as in Chapter 3, the superscript n indicates variables at the "new" time $t + \delta t$, while superscript o indicates variables at the "old" time t. The quantity ε_r is the explicitness parameter for the interface position, $0 \le \varepsilon_r \le 1$. When $\varepsilon_r < 1$, the method is at least partially implicit. When $\varepsilon_r = \frac{1}{2}$, the method is centered and nominally most accurate. If long time steps are contemplated, $\varepsilon_r \le \frac{1}{2}$ is required for Courant stability, with strict inequality usually required to deal with nonlinear effects. When $\varepsilon_r = 0$, the calculation is fully implicit, i.e., fully forward differenced. This choice is most stable but is only first-order accurate. Currently ADINC uses the same value of ε_r at every interface, varying it from cycle to cycle.

The momentum equation (5.16) yields the acceleration of interface i,

$$\frac{dv_i}{dt} = \frac{-1}{\rho_i} \left.\frac{\partial p}{\partial r}\right|_i, \tag{5.24}$$

where the density and pressure gradient are also needed at cell interfaces. The discretization used in ADINC is

$$v_i^n = v_i^o - \frac{\delta t \varepsilon_v}{\langle \rho \delta r \rangle_{i+1/2}}(p_{i+1}^o - p_i^o) - \frac{\delta t(1 - \varepsilon_v)}{\langle \rho \delta r \rangle_{i+1/2}}(p_{i+1}^n - p_i^n). \tag{5.25}$$

Here ε_v is the explicitness parameter for the interface velocity and has the same properties described above for ε_r. The quantities ε_r and ε_v are distinct in ADINC, but no reason has been uncovered to date for using different values in an actual calculation. The interface average indicated as $\langle \rho \delta r \rangle_{i+1/2}$ is both a spatial and temporal average, as described below. Physical considerations are used to define $\langle \rho \delta r \rangle_{i+1/2}$, so the discretization in (5.25) is insensitive to numerical errors arising from large density discontinuities at the interfaces.

ACCELERATION MATCHING ALGORITHM

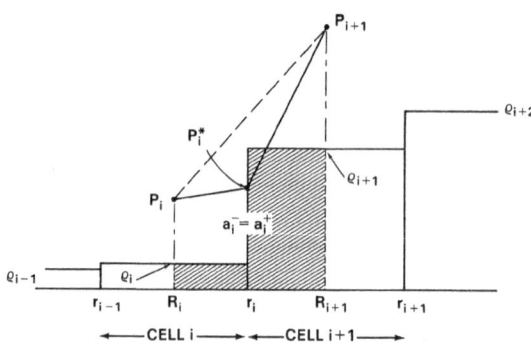

Fig. 5-6 Acceleration-matching technique for density discontinuities at interfaces. An intermediate interface pressure p^* is defined so that the acceleration of material to the right and to the left of the interface is matched. Most of the pressure gradient thus appears across the denser fluid.

Figure 5-6 shows two cells i and $i + 1$ which straddle interface i. The pressures p_i and p_{i+1} are defined at R_i and R_{i+1}, and the densities ρ_i and ρ_{i+1} are assumed constant throughout their respective cells. Because ρ_i and ρ_{i+1} differ spatially (we ignore their time variation for a moment), the straight-line pressure gradient shown would impart different accelerations to the fluid elements just to the right and to the left of interface i. If the fluid were permitted to move according to these distinct accelerations, either the fluids would overlap or a gap would open up at interface i after a short while. To prevent this, a fictional pressure p_i^* is defined at interface i so that the acceleration calculated from the left equals the acceleration calculated from the right:

$$p_i^* = \frac{p_{i+1} f_{i+1}^- + p_i f_i^+}{f_{i+1}^- + f_i^+}, \tag{5.26}$$

where

$$f_i^+ = \frac{1}{\rho_i (r_i - R_i)},$$

$$f_{i+1}^- = \frac{1}{\rho_{i+1}(R_{i+1} - r_i)}. \tag{5.27}$$

By working in terms of the indicated average $\langle \rho \delta r \rangle_{i+1/2}$ in Eq. (5.25), we can eliminate p_i^* completely from further consideration and use

$$\langle \rho \delta r \rangle_{i+1/2} = \rho_{i+1}(R_{i+1} - r_i) + \rho_i(r_i - R_i) \tag{5.28}$$

to define the spatial part of the as yet unspecified average.

The question of how to evaluate the time average of (5.28) has not been fully settled. The major points to consider are momentum conservation, non-linear stability of the overall algorithm, and time-centering accuracy. When multiplied by $\langle \rho \delta r \rangle_{i+1/2}$ and summed over cell boundaries, Eq. (5.25) yields

$$\sum_i v_i^n \langle \rho \delta r \rangle_{i+1/2} = \sum_i v_i^0 \langle \rho \delta r \rangle_{i+1/2} + \text{boundary terms}, \tag{5.29}$$

where the pressure terms cancel except at the boundary of the computational region. One would like to use the old time values on the right and the new time values on the left to give a true momentum integral, at least in Cartesian coordinates. Since the quantity $\langle \rho \delta r \rangle_{i+1/2}$ is conserved in Cartesian coordinates, however, its value is independent of time; it is just the mass associated with interface i. In curvilinear coordinate systems the momentum integral has little meaning and the quantity $\langle \rho \delta r \rangle_{i+1/2}$ is not a constant of the motion; instead $\rho_i \Lambda_i$ is constant. ADINC uses an exactly time-centered average for the geometric parts of $\langle \rho \delta r \rangle_{i+1/2}$. The best approximation (latest iteration) to the new density is used in the expression for reasons of numerical stability.

In Eq. (5.28) ADINC actually uses

$$\langle \rho \delta r \rangle_{i+1/2} = [\rho_{i+1}^P (R_{i+1}^h - r_i^h) + \rho_i^P (r_i^h - R_i^h)], \tag{5.30}$$

where superscript P stands for "previous" and indicates the latest iterated approximation to the new value of the variable, in this case $\{\rho_i^n\}$. The superscript h is used to indicate the exact "half" time average. In Eq. (5.30)

$$r_i^h = \tfrac{1}{2}(r_i^0 + r_i^P) \tag{5.31}$$

and

$$R_i^h = \tfrac{1}{2}(R_i^0 + R_i^P). \tag{5.32}$$

We make no representation that these are the best averages or that extensive testing of this aspect of ADINC has been performed. Perhaps the freedom remaining in this part of the calculation can be used to further improve the accuracy and veracity of the algorithm. No problems traceable to these choices have been observed to date in numerous test calculations using ADINC.

Next we wish to find a tridiagonal equation for $\{p_i^n\}$. The momentum equation (5.25) can be simplified as follows:

$$v_i^n = a_i - b_i(p_{i+1}^n - p_i^n), \qquad i = 2, \ldots, N, \tag{5.33}$$

where

$$a_i = v_i^0 - \frac{\delta t \varepsilon_v}{\langle \rho \delta r \rangle_{i+1/2}} (p_{i+1}^0 - p_i^0),$$

$$b_i = \frac{\delta t (1 - \varepsilon_v)}{\langle \rho \delta r \rangle_{i+1/2}}. \tag{5.34}$$

The equation of state is introduced by requiring that the cell volume Λ_i^{eos} computed from the equation of state using the new time values of pressure equal the new cell volume computed from the fluid dynamics, Λ_i^{fd}. At any iteration the difference is

$$\delta\Lambda_i = \Lambda_i^{\text{eos}}(p_i, s_i, \ldots) - \Lambda_i^{\text{fd}}(\{r_i\}). \tag{5.35}$$

Iteration should be continued until $\delta\Lambda_i$ vanishes. Changing p_i^P to p_i^n varies both terms in Eq. (5.35). In the fluid dynamics contribution, r_i^P converges to r_i^n as a function of the pressure through Eqs. (5.33) and (5.23). We use a Newton–Raphson approach to obtain a quadratically convergent iteration to the desired solution at time $t + \delta t$,

$$\Lambda_i^{\text{eos}}(p_i^n, s_i^n, \cdots) = \Lambda_i^{\text{fd}}(\{r_i^n\}). \tag{5.36}$$

The difference between Λ_i^{fd}, which is known at each iteration, and the desired Λ_i^n, the result of carrying the iteration to convergence, can be written in terms of the cell interface areas and the desired new fluid velocities at the cell interfaces:

$$\Lambda_i^n - \Lambda_i^{\text{fd}} \approx (1 - \varepsilon_r)\delta t[A_i^h(v_i^n - v_i^P) - A_{i-1}^h(v_{i-1}^n - v_{i-1}^P)]. \tag{5.37}$$

The same treatment of the equation of state gives

$$\Lambda_i^n - \Lambda_i^{\text{eos}} \approx (p_i^n - p_i^P)\left(\frac{\partial\Lambda}{\partial p}\right)_i^{\text{eos}}. \tag{5.38}$$

Let

$$
\begin{aligned}
d_i^+ &= -(1 - \varepsilon_r)\delta t A_i^h, \\
d_i^- &= -(1 - \varepsilon_r)\delta t A_{i-1}^h.
\end{aligned} \tag{5.39}
$$

Then Eq. (5.37) becomes

$$\Lambda_i^n - \Lambda_i^{\text{fd}} \approx -d_i^+(v_i^n - v_i^P) + d_i^-(v_{i-1}^n - v_{i-1}^P). \tag{5.40}$$

Equating Λ_i^n in Eqs. (5.38) and (5.40) gives

$$\delta\Lambda_i^P + (p_i^n - p_i^P)\left(\frac{\partial\Lambda}{\partial p}\right)_i^{\text{eos}} \approx -d_i^+(v_i^n - v_i^P) + d_i^-(v_{i-1}^n - v_{i-1}^P). \tag{5.41}$$

Define

$$c_i = p_i^P\left(\frac{\partial\Lambda}{\partial p}\right)_i^{\text{eos}} + d_i^+ v_i^P - d_i^- v_{i-1}^P - \delta\Lambda_i^P. \tag{5.42}$$

Then Eq. (5.41) becomes

$$p_i^n\left(\frac{\partial\Lambda}{\partial p}\right)_i^{\text{eos}} + d_i^+ v_i^n - d_i^- v_{i-1}^n \approx c_i. \tag{5.43}$$

This becomes a tridiagonal implicit linear equation for the estimated new pressures $\{p_i^n\}$ when Eq. (5.33) is used to express $\{v_i^n\}$ in terms of $\{p_i^n\}$. Expanding (5.43) gives

$$p_i^n \left(\frac{\partial \Lambda}{\partial p}\right)_i^{eos} + d_i^+ \left[a_i - b_i(p_{i+1}^n - p_i^n)\right]$$

$$- d_i^- \left[a_{i-1} - b_{i-1}(p_i^n - p_{i-1}^n)\right] \approx c_i. \tag{5.44}$$

Equation (5.44) is the basic equation solved by ADINC. Iteration is necessary because Eqs. (5.37) and (5.38) hold only to first order. The iteration is quadratically convergent, however, because only second-order terms are neglected at each stage of the iteration. For the interior cells $i = 3, 4, \ldots, N$, Eq. (5.44) may be written in the form

$$A_i^* p_{i-1}^n + B_i^* p_i^n + C_i^* p_{i+1}^n = D_i^*. \tag{5.45}$$

At the right and left boundaries of the system the new velocities in (5.42) and (5.43) are imposed externally. In terms of the tridiagonal coefficients of Eq. (5.45) we have

$$A_2^* = C_{N+1}^* = 0. \tag{5.46}$$

When integrating the fluid dynamic equations, ADINC assumes that each interface moves in a fully Lagrangian manner according to Eq. (5.22). The change in density from one time step to the next in a cell is therefore given simply by the change in cell volume according to the mass conservation equation

$$\rho_i^n \Lambda_i^n = \Delta M_i = \rho_i^o \Lambda_i^o, \tag{5.47}$$

When individual species number densities must be followed, they are also advanced by Eq. (5.47).

If we define

$$\beta = \frac{2c_s \delta t}{\delta x} \sin\left(\frac{1}{2}k\delta x\right), \tag{5.48}$$

where c_s is the sound speed, it is possible to show that a small-amplitude sinusoidal wave approximated according to the above algorithm is multiplied by a factor

$$\frac{\rho_i^n}{\rho_i^o} = \frac{1 - \frac{1}{2}\beta^2 \left[\varepsilon_v(1-\varepsilon_r) + \varepsilon_r(1-\varepsilon_v)\right] \pm i\beta \left\{1 - \frac{1}{4}\beta^2 \left[\varepsilon_v(1-\varepsilon_r) - \varepsilon_r(1-\varepsilon_v)\right]^2\right\}^{1/2}}{1 + \beta^2(1-\varepsilon_r)(1-\varepsilon_v)} \tag{5.49}$$

per time step. The absolute value $\mathscr{A}$ of the amplification satisfies

$$\mathscr{A}^2 = \frac{1 + \beta^2 \varepsilon_r \varepsilon_v}{1 + \beta^2(1-\varepsilon_r)(1-\varepsilon_v)}. \tag{5.50}$$

To measure wave damping, a series of calculations was performed using 20 cells per wavelength and 20 time steps per cycle as the standard conditions. Figure 5-7 summarizes a number of computations performed with different

WAVE DAMPING IN ADINC

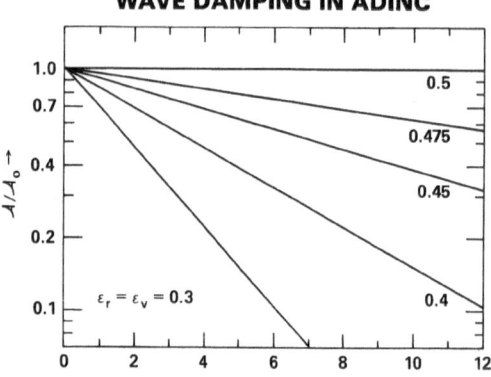

Fig. 5-7 Wave amplitude as a function of time, illustrating damping introduced by forward differencing. Amplitudes are measured in units of the initial amplitude. The values of the explicitness parameters are shown. With $\varepsilon_r = \varepsilon_v = 0.45$, for example, the wave has damped by a factor greater than 2.5 in 10 wave periods.

values of the explicitness parameters ε_r and ε_v. The wave amplitude relative to the initial wave amplitude is plotted versus the time measured in wave periods. Here the amplitude is determined by subtracting from the total energy the thermal energy the system would have with zero velocity. The resulting linearity versus t/τ on a semilogarithmic scale demonstrates that the numerical damping is exponential, as expected. Even with the very modest damping introduced by $\varepsilon_r = \varepsilon_v = 0.45$, the sound wave decays by more than a factor of 2.5 in only 10 oscillation periods. This means that the use of fully forward-differenced schemes for numerical stability makes the meaningful simulation of sound waves essentially impossible. Calculations should be carried out with ε_r, ε_v as close to 0.5 as possible, so that accuracy can be maintained as well as stability.

Any discussion of the accuracy and stability of ADINC presupposes the monotonicity of the interface positions $\{r_i\}$ and, of course, the positivity of the interface areas $\{A_i\}$. Since the algorithm is based on a discretization of the continuum fluid dynamic equations, numerical error and even instability can result from nonphysical crossing of cell interfaces, even though the implicit algorithm given above is nominally stable for sound waves at arbitrary time step. To prevent interface crossings, a Courant condition must still be satisfied for the flow velocities $\{v_i\}$ even though $|v| \ll c_s = (\partial p/\partial \rho)^{1/2}$ throughout the fluid. In the reference version of ADINC reproduced in Appendix D, the maximum time step which still prevents interfaces from crossing adjacent interface positions is calculated from the formula

$$\delta t_{max} = \frac{1}{2} \min_i \left\{ \frac{r_{i+1} - r_i}{|v_i|}, \frac{r_i - r_{i-1}}{|v_i|} \right\}, \tag{5.51}$$

where a very small number is added to $|v_i|$ to prevent division by zero.

Equation (5.51) is generally conservative and even includes the factor of $\frac{1}{2}$ in case two interfaces are moving toward each other. By rights only one of the two terms should be included since an interface can be moving only to the right or to the left. Furthermore, interfaces cross only due to a differential velocity, not an absolute one. The denominators in Eq. (5.51) should be $|v_{i+1} - v_i|$ and $|v_i - v_{i-1}|$ when a net motion is superimposed on relative expansions and contractions.

Situations exist, however, where the time step calculated from Eq. (5.51) can lead to trouble. From Eq. (5.23) one can see that the cell interface positions are advanced using an average of the new and old velocities. Since the time step has to be estimated at the beginning of a cycle, it is quite possible for $\{v_i^o\}$ to be small or zero while $\{v_i^n\}$ is large. Boris (1979) discusses how to choose the time step so as to avoid this possibility and also the convergence criterion used to terminate the pressure iteration. Both of these depend to some degree on the nature of the problem being solved, and thus invite the intervention of the user.

The version of the ADINC algorithm implemented on the NRL ASC is listed in Appendix C. A user applies it via a series of structured calls to the nine entries in the three major routines dealing with geometry, the equation of state, and the fluid dynamics. It should be clear that the iterative procedure on which the algorithm is based does not require this particular structure and is, in fact, applicable to a broad class of Lagrangian codes.

5.4 Examples

The problem which originally motivated the development of ADINC was that of simulating the behavior of the slightly compressible liquid liners used in the Linus experiments (Book et al., 1975, 1977). Linus uses a magnetically, mechanically, or gas-propelled imploding liquid liner to compress a charge or "payload" consisting of plasma and magnetic flux. The earliest analyses of the system assumed the liner to be a constant-density incompressible ideal (inviscid) liquid, However, liner compressibility can actually have a substantial effect on the dynamics, energetics, and dwell time of the implosion by altering the minimum payload volume and timing of the liner motion (Book and Turchi, 1979). To study these effects it was necessary to simulate acoustic phenomena accurately for time periods long compared with the sonic transit time t_s.

In the Helius and Linus-0 experiments, a piston is displaced axially by the pressure of gas in a surrounding plenum. The piston forces a liquid liner (water or sodium–potassium eutectic mixture) through ducts which constrain it to implode radially. The axial and radial motions are superposed on the azimuthal motion resulting from the rotation of the container, which imposes axisymmetry. To describe this behavior quantitatively requires knowledge of the driving pressure and duct geometry, and inclusion of angular momentum

THEORETICAL TRAJECTORY

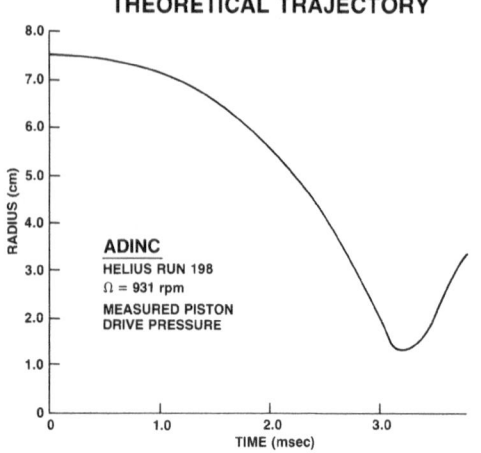

Fig. 5-8 Result of using ADINC to calculate liner trajectory (inner radius as function of time) with measured piston drive pressures as boundary conditions.

conservation and a realistic equation of state in the radial and axial momentum equations. For many purposes the latter equation can be omitted, provided some approximation is used to include the effects of geometry.

The driving pressure is measured experimentally. In some versions of the experiments the liner inner radius is directly observable; in others, its position is inferred from probe measurements of the compression of a trace or "seed" axial magnetic field.

The liner, which starts from rest, accelerates until the gas or plasma–magnetic field core becomes highly compressed. When the pressure interior to the liner is high enough, the implosion reverses and is converted to an accelerating expansion. At "turnaround" the inner surface of the liner comes to rest. At this point, the time step is large because the fluid is moving slowly, even though the accelerations are maximum. To avoid the problem of taking too large a time step in the test program, both the old velocity and the current velocity are used to get the smallest estimate of δt available. Then the estimated δt is limited to be no more than 1.1 times the previous time step. This last operation ensures that the step does not automatically become large near turnaround.

Figure 5-8 displays the calculated position of the inner edge of the liner as a function of time near turnaround (Cooper *et al.*, 1980). Some asymmetry seems evident, connected with the weak compression of the liner material near turnaround. The driving pressures were obtained from a fit to experimental measurements. In Fig. 5-9 the value of the time rate of change of the seed field corresponding to this radial trajectory is compared with the measured values. Note that when corrections are made for "fringing" of the magnetic field near the ends of the system (another geometrical effect which requires a two-dimensional treatment for detailed modeling), the measured and predicted values of $\dot{B}$ are in close agreement.

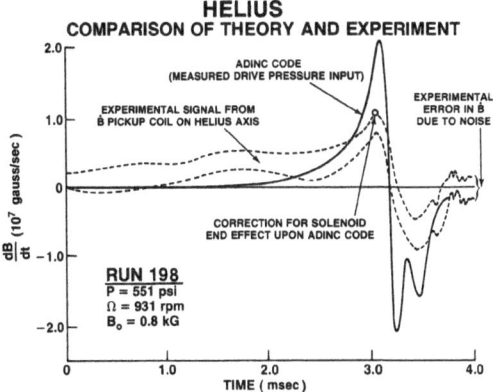

Fig. 5-9 Comparison of measured strength of seed magnetic flux, compressed in order to provide a diagnostic of the liner trajectory (dashed curves) with calculated values corresponding to the trajectory of Fig. 5-8 (solid curve). The circle denotes the location of the calculated peak after the latter is corrected for field fringing at the ends of the device.

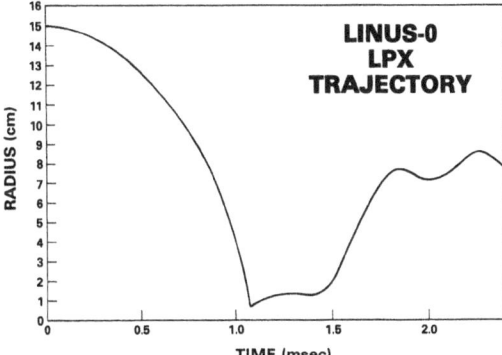

Fig. 5-10 Calculated Linus-0 trajectory, showing pronounced effects of liner compressibility. The extended dwell period (flattened region near minimum) is accompanied by negative pressures (the liner is under tension during part of the rebound). The ripples on the right are produced by radially propagating sound waves.

Figure 5-10 shows a prediction of a liner trajectory for the larger Linus-0 device. Here the driving pressure (340 atm) is roughly 10 times that used in the previous case. The asymmetry in liner motion is now quite pronounced. Because the outer portion of the liner continues to implode for about $\frac{1}{2}$ ms after the inner surface turns around, the period of maximum compression has been extended by more than a factor of 10 as a consequence of the compression taking place there.

As a second example of calculations carried out using ADINC, we present results of flame ignition studies made by Oran, Young, and Boris (1980). Ignition of fuel–oxidizer mixtures occurs when an external source of energy initiates interactions among the controlling convective, transport, and chemical processes. Whether the process is quenched, deflagrates, or detonates depends on the energy intensity, duration, and volume affected. It also depends on the initial properties of the mixture, which determine the chemical induction time and the heat release per gram of material. Ignition is a complicated phenomenon whose occurrence for a specific mixture of fuel and oxidizer depends strongly on diffusive and chemical parameters which are often very poorly known. A convenient, inexpensive way to estimate whether a given heat source will ignite a mixture would be both a valuable laboratory tool and a useful learning device.

The model consists of coupled conservation equations for the densities of total mass ρ, momentum $\rho\mathbf{v}$, and energy $\mathscr{E}$, as well as the individual species densities $\{n_j\}$:

$$\frac{\partial \rho}{\partial t} = -\nabla \cdot \rho\mathbf{v}, \tag{5.52}$$

$$\frac{\partial n_j}{\partial t} = -\nabla \cdot n_j\mathbf{v}_j - \nabla \cdot n_j\mathbf{v} + P_j - Q_j n_j, \tag{5.53}$$

$$\frac{\partial \rho\mathbf{v}}{\partial t} = -\nabla \cdot (\rho\mathbf{v}\mathbf{v}) - \nabla\mathbf{p} + \nabla \cdot \eta_m(\nabla\mathbf{v} + (\nabla\mathbf{v})^T), \tag{5.54}$$

$$\frac{\partial \mathscr{E}}{\partial t} = -\nabla \cdot \mathscr{E}\mathbf{v} - \nabla \cdot (p\mathbf{v} - \lambda_m\nabla T - Q_D). \tag{5.55}$$

The quantity $\mathbf{v}$ is the fluid velocity, the $\{\mathbf{v}_j\}$ are the diffusion velocities, and $\{P_j\}$ and $\{Q_j\}$ refer to the chemical production and loss processes for the individual species j, respectively. The quantities η_m and λ_m are the mixture viscosities or thermal conductivities. The superscript T in the last term of Eq. (5.54) indicates that the transpose is to be taken. The quantity Q_D, representing the contributions to the thermal conductivity from diffusion processes, is a function of $\{\mathbf{v}_j\}$, the binary molecular diffusion coefficients $\{D_{jk}\}$, the thermal diffusion coefficients $\{D_j^T\}$, and the enthalpies $\{h_j\}$. Solution of the diffusion velocities requires treating the binary and thermal diffusion terms and the pressure gradient term, which in this flame is negligible. For mixtures of ideal gases, the pressure satisfies $p = nk_BT$, where n is the total number density, k_B is Boltzmann's constant, and T is the temperature. It is assumed that there are no radiative losses in the system.

The technique for solving (5.52)–(5.55) is based on the method of asymptotic time-step splitting. The individual chemical and physical processes are solved separately by the fastest and most accurate algorithms and then asymptotically coupled together. The chemical interactions are described by a set of nonlinear coupled ordinary differential equations. These are integrated using VSAIM, a fully vectorized version of the selected asymptotic integration method CHEMEQ developed by Young and Boris (1977). The convective transport is solved using ADINC. Tests of the fully detailed flame model in the limit of no chemical reactions or diffusive transport are essentially tests of the ADINC algorithm. A number of such tests have been documented by Boris (1979). When vectorized for the NRL ASC, the code requires between 0.05 and 0.08 s per time step to evaluate the full set of equations for 45 cells.

Results from a typical flame calculation in Cartesian geometry are shown in Figs. 5-11, 5-12, and 5-13 (Oran and Boris, 1979). Figure 5-11 shows a typical

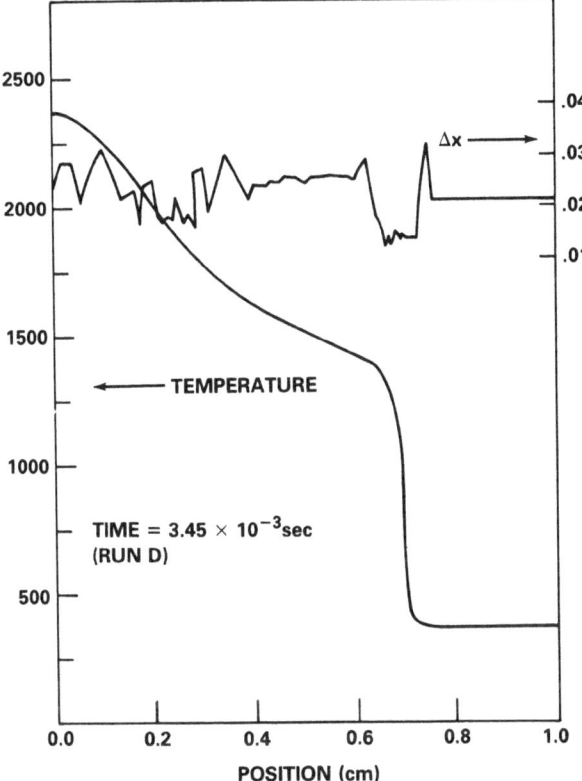

Fig. 5-11 Temperature profile calculated from the detailed simulation model for a flame in an $H_2 : O_2 : N_2 = 2 : 1 : 10$ mixture ignited by a quadratic temperature profile at the onset of the calculation. Also shown is a graph of the cell size, Δx, as a function of position.

temperature profile for a flame initiated by a quadratic temperature distribution at the onset of the calculation. Also drawn on the same figure is a plot of the cell size Δx as a function of position, showing that the resolution decreases as the absolute magnitude of temperature gradient increases. It is clear that the considerable variation in cell size introduces no related variation in the temperature, as would occur if the results of the convective transport depended strongly on the local value of Δx. Species profiles are shown in Fig. 5-12 and details of the flame front in Fig. 5-13. We note that the intermediate species densities and the temperature gradients do not all peak at the same location, and the structure of the flame front is clearly distinguishable.

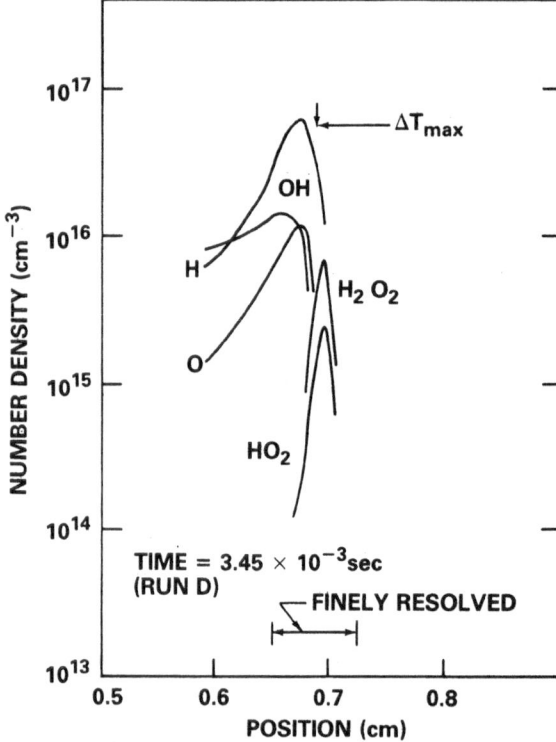

Fig. 5-12 Calculated profiles of species concentrations corresponding to the temperature profile in Fig. 5-11.

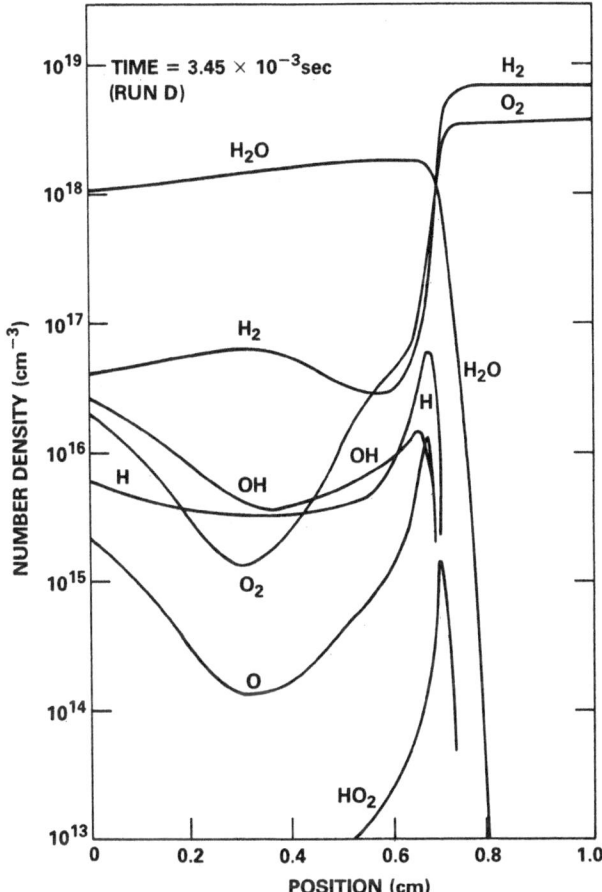

Fig. 5-13 Calculated profiles of intermediate species concentrations at the flame front corresponding to Figs. 5-11 and 5-12. The fine resolution around the flame front allows differentiation of the peak densities of the intermediate species.

CHAPTER 6

Two-Dimensional Lagrangian Fluid Dynamics Using Triangular Grids

6.1 Grid Distortion in Two Dimensions

Section 5.1 stressed the deleterious effects of mesh distortion on Lagrangian methods. The loss in accuracy due to mesh distortion applies, of course, to higher dimensions as well. In practice, Lagrangian techniques have been limited to one-dimensional calculations or flows in higher dimensions which are very "well behaved," since shear, fluid separation, and large-amplitude motions produce severe grid distortions. The mesh points commonly used to evaluate gradients and Laplacians are shown in Fig. 6-1a for a regular two-dimensional grid. Figure 6-1b illustrates a simple grid distortion produced by shear flow. A well-formulated Lagrangian finite-difference algorithm will properly account for the angle between grid lines and the variable mesh spacing produced by this distortion. Nevertheless, numerical approximations based on this mesh can still be grossly in error because differences no longer involve neighboring vertices. Mesh points which are now closer to the central vertex do not enter into the approximation, while those farther away do.

Higher-order approximations may lead to even greater error. Figure 6-2a shows the vertices commonly used in higher-order approximations. Figure 6-2b shows how these approximations on a distorted mesh may include vertices which are even further removed from the central vertex, while neglecting other vertices which lie closer. In other words, the distorted mesh cannot be used to self-consistently improve the accuracy of the approximation. The problem can be resolved only by ensuring that differencing is carried out over the appropriate vertices. The mesh distortion must be reduced.

The conventional solution to the problem of distortion has been to maintain a regular mesh by allowing fluid to pass through the mesh during a rezoning phase. Figure 6-3 illustrates rezoning for a mesh about a shear layer. A

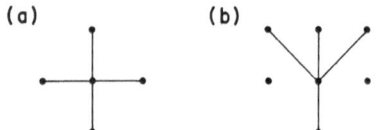

(a) (b)

Fig. 6-1 (a) Five-point stencil used in finite-difference approximation to, e.g., Laplacian. (b) Lagrangian motion of points distorts the grid, reducing the accuracy of the approximation.

Lagrangian calculation very quickly leads to the mesh shown. Eulerian rezoning can be used to prevent further distortion, but for the rest of the calculation the shear interface is effectively Eulerian. At first the portion of the grid immediately adjacent to the shear region remains Lagrangian, but subsequently it moves relative to the now Eulerian grid at the shear interface. Soon it, too, distorts and allows Eulerian slippage. In this manner the whole grid may become Eulerian in the presence of shear. Unless the algorithms used to achieve slippage are as accurate as any that could be used to difference the nonlinear advective terms in a purely Eulerian calculation, at later times this "Lagrangian" calculation must be less accurate than an Eulerian one. In short, the rezoning technique may introduce gross error through a low-accuracy rezoning algorithm.

A related but separate question arises from the effect of rezoning even one vertex. For example, on a quadrilateral mesh a central vertex may be moved to the average position of its neighbors. The Eulerian motion of the vertex causes fluid to pass through the four quadrilateral cell sides meeting at that vertex, affecting the physical quantities specified in all four cells and at the four neighboring vertices as well as the central vertex. The result of the rezone phase is therefore the introduction of artificial diffusion over one grid spacing in any direction.

A third problem is that Eulerian slippage is difficult to employ in some flows, such as those which occur in an initially simply connected region which evolves into a multiply connected region. Finally, the most serious problem is that use of Eulerian techniques tends to obscure the basic question of what is the best approximation possible on a given Lagrangian grid. The simplest rezoning techniques, e.g., averaging the nearest-neighbor positions to determine the new central vertex position, may not significantly reduce the error

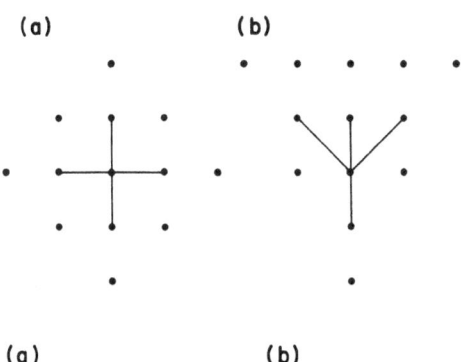

Fig. 6-2 As in Fig. 6-1, but for nine-point stencils. Central point and connections to nearest neighbors are shown (a) before and (b) after Lagrangian grid point motion.

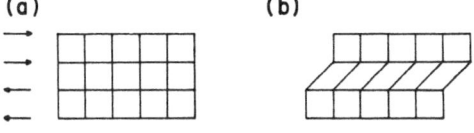

Fig. 6-3 Rezoning applied to mesh in the neighborhood of shear flow (arrows). Initially rectangular cells (a) are elongated and distorted by the flow (b).

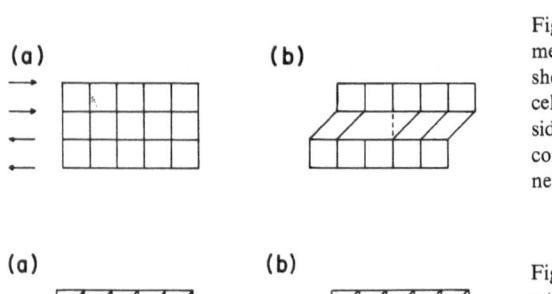

Fig. 6-4 Reconnection applied to mesh in the neighborhood of shear flow. Initially rectangular cells (a) can become three- or five-sided (b) as a result of new connections drawn between neighboring vertices.

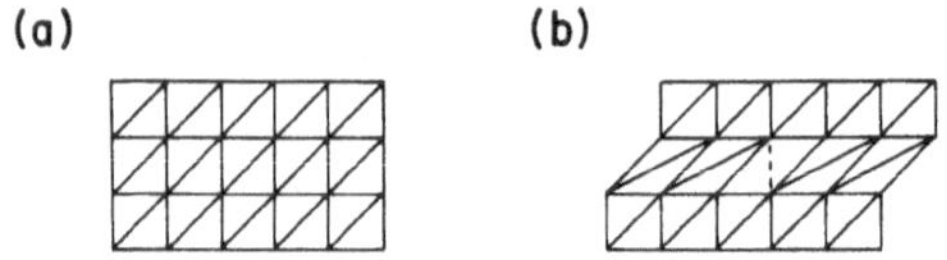

Fig. 6-5 Reconnection applied to a triangular grid (a) in the neighborhood of shear flow leaves the grid triangular (b).

for a distorted mesh. In a uniform shear flow the central vertex position is exactly the average position of the original neighbors, regardless of the distortion. In other words, rezoning techniques may preserve a grid which forces an inaccurate representation.

6.2 Use of Reconnection to Eliminate Grid Distortion

Rezoning attempts to solve the problem of grid deformation by using vertex motion to rectify the distortions. An alternate solution is illustrated in Fig. 6-4. A section of a quadrilateral mesh about a shear layer is shown in Fig. 6-4a. A Lagrangian calculation quickly leads to the mesh shown in Fig. 6-4b, in which mesh connections about the layer no longer join neighboring vertices. In this figure one grid line has been relocated to show the connection which is now appropriate. For a periodic system all stretched grid lines could be reconnected, thereby restoring the mesh to its original configuration without relocating any vertices. In general, some reconnections either are inappropriate or cannot be made due to boundaries, so that one triangular and one pentagonal cell remain, as shown in Fig. 6-4b. Therefore reconnection on a quadrilateral grid cannot totally resolve the problem of distorted grids.

On a triangular mesh, however, there are no such complications. As shown in Fig. 6-5, a reconnected grid line on a triangular mesh still results in two triangular cells. This technique, first used in a computer code by Crowley (1971), represents a very attractive alternative to rezoning for triangular grids. There is no Eulerian vertex motion. For a given reconnection only one grid line or two cells are affected, instead of all the grid lines and cells about a given vertex as in the rezone technique. Therefore, diffusion takes place effectively over at most one cell, instead of one in either direction.

The major similarity between reconnection and rezoning is that by itself neither can treat fluids evolving into multiply connected regions or flows at stagnation points. Reconnection does, however, help somewhat. The number of grid lines meeting at any vertex can be reduced to three by reconnections, with the three neighboring vertices forming a triangle which surrounds only that vertex. If the fluid is accumulating too many vertices locally, then that central vertex and its three grid lines can be eliminated, leaving only the surrounding triangle. This produces the desired decrease in resolution and avoids the formation of long, narrow triangles near the point of converging flow. Conversely, a vertex may be added inside any triangle or along any line by providing the necessary grid lines to other vertices within the affected triangles. Subsequent reconnections will link the added vertices to their neighbors. In this way the transition to multiply connected regions and the flow near stagnation points can be handled smoothly merely by decreasing or increasing resolution where appropriate. The combination of grid line reconnection with vertex addition and deletion therefore provides a means of smoothly restructuring the grid without recourse to Eulerian vertex movement.

The simplicity of the algorithms can be deceiving. Although grid reconnection, vertex deletion, and vertex addition seem trivial by themselves, the combined effect of the technique is to render soluble a large class of previously intractable computational problems in fluid dynamics. A good example of the type of problem requiring improved numerical techniques is the problem of a large-amplitude breaking surface wave. When the wave top separates from the wave due to bottom shallowing, nonlinear wave interaction, or strong surface winds, it falls back into the wave trough in a complicated flow which the usual techniques cannot handle. The Lagrangian triangular grid algorithms discussed above allow the crest of the wave to separate from the body of the wave, fall under the influence of gravity and any strong winds back into the trough, and become reabsorbed smoothly, while following the numerical representation of the fluid at the surface. Figure 6-6 illustrates a Lagrangian triangular mesh used to solve this problem. As the neck of fluid narrows, the corresponding triangle sides shrink to zero and the connection between the two bodies of dense fluid is broken. In this way the topology of the problem changes self-consistently with the evolution of the system (Boris, Fritts, and Hain, 1975).

A triangular-element mesh has several other advantages. Since the number of triangles meeting at a vertex is variable, increased accuracy in one region of the flow does not force unnecessary resolution in other areas of the flow. This versatility also permits both regular and irregular tessellations of the x–y-plane with triangles. Triangles, unlike rectangles, can cover a circle symmetrically without cusps or other local representation irregularities. Free surfaces, complicated interfaces, sharp gradients, and boundaries of immersed objects can be represented accurately and economically.

The triangle is a much less ambiguous structure than a rectangle or higher-

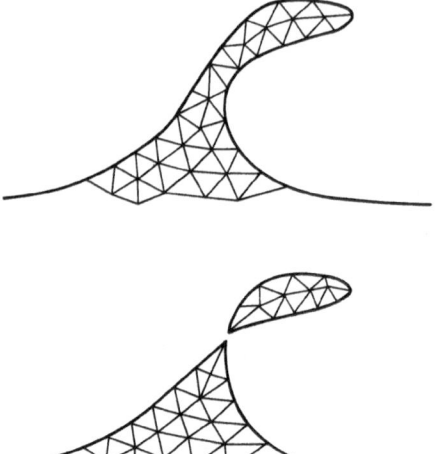

Fig. 6-6 Schematic illustration of the use of a two-dimensional grid of triangles to represent a large-amplitude, separating, free-surface gravity wave.

order polygon, and hence interpolations and integrals are usually simpler to perform. Because the figures are three sided (rather than four sided), there is no unambiguous labeling of grid points as even and odd. Consequently, the even–odd computational instability of rectangular schemes appears to be absent, or at least greatly subdued, in the triangular representation. A vertex which is one point removed from its neighbor along a particular path will be two points removed by another. This does not mean that an even–odd problem cannot occur, only that it is not a topologically natural mode of instability and hence is considerably slower growing than its rectangular-mesh counterpart.

Of course, there are problems using the triangles. An irregular triangular grid with symmetric connections imposes a greater calculational overhead because of the varying lengths of lines within the grid. If the grid, furthermore, is allowed to have an arbitrary number of lines meeting at a vertex (a general connectivity mesh), special bookkeeping arrays must be instituted to record all important aspects of the mesh interconnection. When each vertex, each side, and each triangle are numbered, records of interconnections simply take the form of an ordered series of integers, which can be stored compactly in the computer. For example, the information might be stored in the form of lists: vertex lists enumerating the sides and triangles about any vertex; side lists enumerating the vertices starting and finishing the side and the adjacent triangles; and triangle lists containing the sides and vertices defining the triangles. Since these lists are not all independent, they are not all needed. Maintaining some redundant lists is costly in terms of storage, but there are real advantages gained in increased computation speed, clarity, and ease of coding.

If, in addition, the general connectivity mesh is allowed to reconnect and add and subtract vertices, adjacent elements in arrays cease to belong to adjacent triangles. It is impossible to sweep through the mesh in a systematic fashion, and as a result, finite-difference algorithms are harder to vectorize. Thus, the price paid for flexibility is the loss of global ordering. The lack of a global representation in which the vertex numbering corresponds to spatial position also greatly complicates implicit calculations. Poisson's equation, for example, is solved by iteration, and explicit time and space derivatives are the best that one has any right to hope for. Partly because of this difficulty, general triangular grides have not received the attention given to regular grids and are not as well understood. The relative lack of experience in differencing over general triangular grids has historically led to mistakes in difference algorithms. The resulting poor behavior of computer codes using these faulty algorithms has been translated into fairly widespread myths about the "stiff" behavior of any scheme employing triangular grids.

The attribution of stiffness to triangular grids seems to have evolved from an improper placement of physical variables on the mesh. In a quadrilateral mesh, there is a one-to-one correspondence between cells and vertices. Thus, for example, each quadrilateral cell may be associated with its upper right vertex. The remaining vertices lie on boundaries, and their variables are specified by boundary conditions. If that same quadrilateral mesh is subdivided into a triangular mesh by drawing diagonals through each cell, the correspondence between cells and vertices is clearly destroyed. The number of vertices is the same, but there are twice the number of cells. On the quadrilateral mesh it is attractive to assign pressures as cell-centered quantities since this makes pressure forces easy to calculate. Such a positioning on a triangular mesh can be disastrous. Twice as many pressures must be specified as in the well-formulated quadrilateral case. Numerically, an iterative solution for pressures would converge extremely slowly, and in this sense the system of differenced equations would be stiff.

If the idea of conserving cell size for incompressible flows is also carried over to triangular grids, the algorithms are stiff in a much worse sense. A vortex is shown in Fig. 6-7, where the arrows designate local fluid velocities. All velocities are about the vortex center, and there exists a maximum speed at some radial distance from the center. The triangle shown in Fig. 6-7 will therefore eventually evert since the most rapidly moving vertex must pass between the two more slowly moving vertices. If an attempt is made to conserve triangle areas, a pressure must build up to resist eversion until the speeds of all vertices are consistent with cell area conservation. The presence of such a pressure component at any stage of the calculation is totally nonphysical; it arises from improper placement of variables and is directly opposed to the physical flow. Such behavior is definitely "stiff" since the solution will tend to resemble solid body rotation rather than vortex motion. In short, stiffness in triangular grids seems to be due entirely to such improper differencing.

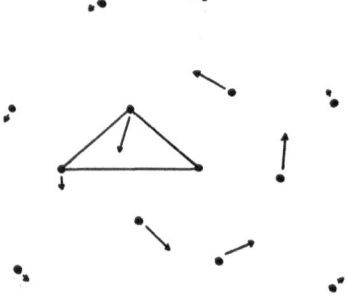

Fig. 6-7 Attempting to conserve cell size
(triangle areas) in a triangular grid can lead to
nonphysical constraints on Lagrangian grid
motion. The triangle shown must evert in the
vortex-induced flow. Conserving cell size alters
the correct Lagrangian velocities.

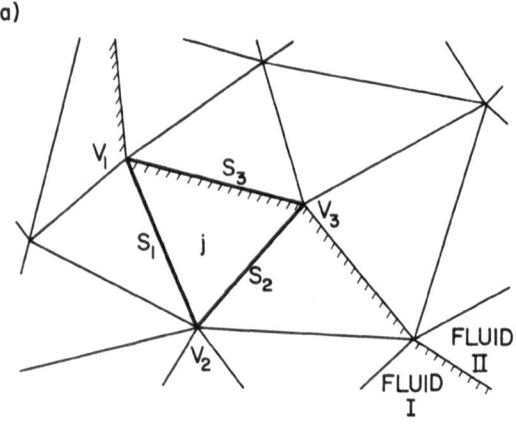

Fig. 6-8 (a) sides and
vertices of a triangle j
located at the interface
between fluid I and fluid
II; (b) the same, with the
bisectors of the sides
drawn to define the cell
associated with vertex V_3.

Although the vortex problem shows that triangle-centered pressures lead to stiff behavior, even with proper placement of these variables, the resulting finite-difference algorithms may still be stiff. We have discovered a topological constraint on the number of vertices attached to a given vertex (Fritts and Boris, 1980), which further restricts the means for implementing conservation of vorticity and mass. We have shown that if velocities are specified on vertices, the grid structure can cause the conservation of mass and of vorticity to be incompatible locally. With exact vorticity conservation numerical fluctuations in divergence arise near vertices having more than six connections. It is possible to alleviate this problem through a definition of vorticity based on triangle-centered velocities, however, which leads to exact conservation schemes for vorticity with no concomitant divergence fluctuations.

6.3 Numerical Algorithms

Figure 6-8 shows a section of a triangular mesh representation with an interface between two fluids of types I and II. The basic elements involved in the construction of a triangular mesh are shown. In Fig. 6-8a a particular triangle j is shown in heavy lines and the various components of that triangle are labeled. Three vertices, V_1, V_2, and V_3, are connected consecutively by sides S_1, S_2, and S_3. The direction of labeling around each triangle is counter-clockwise and the z-axis is directed out of the page. Since the mesh can be irregularly connected, an arbitrary number of triangles can meet at each vertex. The number of triangles and sides meeting at a vertex is equal, except near free surfaces and boundary surfaces where the grid terminates. Each side (except on boundaries) is bounded by two triangles, and each triangle shares sides with three other triangles.

Figure 6-8b illustrates several important properties of triangles which are used in constructing finite-difference algorithms. It is convenient to define a cell surrounding a vertex, as shown by the shaded region surrounding V_3. The borders of such vertex-centered cells are determined by constructing all of the side bisectors for each triangle. Since the three side bisectors all intersect at a point, as shown for triangle j, there is no ambiguity in constructing the vertex cells as indicated. The point of intersection of the side bisectors, the centroid of the triangle, is the center of gravity for the figure; this is true in r–z-coordinates as well as Cartesian (x–y) coordinates. The three side bisectors of any triangle divide the triangle into six subtriangles. These six subtriangles all have the same area and, therefore, each of the three vertex cells receives one-third of the area of the triangle. When a physical variable is constant over a triangle, the contributions of that variable from a triangle to each of the three vertex cells are also equal. These properties of side bisectors make the calculation of cell volumes and cell masses particularly simple.

The use of pressures defined on vertices permits a definition of pressure gradients in direct analogy with Eq. (5.12). Furthermore, we can rewrite Eq. (5.12) using Eqs. (5.3) and (5.4) for the forward and backward approximations:

$$\frac{\partial p}{\partial x_i} = \frac{(\partial p/\partial x_i)^+ \, \Delta x_i + (\partial p/\partial x_i)^- \, \Delta x_i'}{\Delta x_i + \Delta x_i'}. \tag{6.1}$$

That is, the central difference can be obtained by an area-weighted sum of the forward and backward differences.

This result carries over directly to general triangular grids in two dimensions. We can define the pressure gradient there as

$$\nabla p = \sum_{i=1}^{3} p_i \frac{\hat{\mathbf{z}} \times (\mathbf{r}_{i-1} - \mathbf{r}_{i+1})}{2A}. \tag{6.2}$$

The right-hand side of (6.2) is the first-order-accurate finite-difference approximation to the gradient evaluated at the triangle centroid. Here A is the triangle area, $\hat{\mathbf{z}}$ is the unit vector in the direction of the ignorable coordinate, and the sum extends over the three triangle vertices. The analog of Eq. (6.1) is

$$\nabla p_i = \frac{\sum_j A_j \nabla p_j}{\sum_j A_j}, \tag{6.3}$$

where the index j labels triangle-centered quantities and i labels vertices. For special geometries Eq. (6.3) is second-order accurate or higher. In most situations it is less accurate, but it yields accuracy reasonably close to second order for a general mesh, provided that the triangle areas are nearly equal. In that case the error is determined by a formula similar to Eq. (5.11). The worst that can be achieved is first-order accuracy, which occurs only in the degenerate case of a zero-area triangle. This implies that care must be taken in evaluating boundary conditions, but this task is alleviated by having the pressures specified at the boundaries. In a cell-centered scheme, the pressures are located half a cell away, and boundary conditions, particularly at free surfaces, are more difficult to implement. Accuracy in the interior of the fluid is diminished primarily by narrow triangles. As shown below, this restriction is not too serious for a reconnecting grid since the grid can be made to reconnect to preserve regular triangles. Where this is not possible (near interfaces, for example), the addition or deletion of vertices can be used to regularize the mesh.

The conservation integral approach and definitions of divergence we employ allow a natural treatment of compressible systems. We have restricted ourselves, however, to the study of fluids which are inviscid and incompressible, although they may have variable density. We also concentrate attention on problems in which gravity is constant and directed in the negative y-direction.

Though not necessary, these restrictions simplify the analysis and allow the full spectrum of problems of current interest to be solved. The basic equations of the system are Eqs. (2.1)–(2.3).

The fluid density ρ, pressure p, and velocity $\mathbf{v}$ are assumed to vary only with x and y. Equation (2.2), the condition for incompressibility, removes the sound waves. We will assume that p is a constant along free surfaces.

With pressures specified at the vertices, ∇p is evaluated over triangles, and Eq. (2.3) can easily be advanced implicitly or explicitly if velocities are considered to be triangle centered. In what follows we will continue to use the subscript i to denote a vertex-centered quantity and j to denote a triangle-centered quantity. The integration of velocities uses a split-step algorithm in which the velocities are advanced one-half time step, the grid is advanced a full time step, and then the velocities advanced forward the other half time step:

$$\mathbf{v}_j^{1/2} = \mathbf{v}_j^o - \frac{\delta t}{2\rho_j}(\nabla p)_j^o - \frac{\delta t}{2}g\hat{\mathbf{y}}, \tag{6.4}$$

$$\mathbf{v}_i^{1/2} = \tfrac{1}{2}(\mathbf{v}_i^o + \mathbf{v}_i^n), \tag{6.5}$$

$$\mathbf{x}_i^n = \mathbf{x}_i^o + \delta t \mathbf{v}_i^{1/2}, \tag{6.6}$$

$$\mathbf{v}_j^{1/2} = \mathfrak{R}(\{\mathbf{x}_i^o\}, \{\mathbf{x}_i^n\}) \cdot \mathbf{v}_j^{1/2}, \tag{6.7}$$

$$\mathbf{v}_j^n = \mathbf{v}_j^{1/2} - \frac{\delta t}{2\rho_j}(\nabla p)_j^n - \frac{\delta t}{2}g\hat{\mathbf{y}}. \tag{6.8}$$

The vertex velocity $\mathbf{v}_i^n$ appearing in Eq. (6.5) is obtained from the area-weighted $\mathbf{v}_j^n$ from the previous iteration,

$$\mathbf{v}_j^n = \frac{\sum\limits_j \mathbf{v}_j^n A_j}{\sum\limits_j A_j}. \tag{6.9}$$

The advantage of using triangle-centered velocities is the ease in conceptualizing and expressing conservation laws. Because of the paucity of experience in formulating algorithms over a general triangular grid, we have employed the control volume approach (cf. Section 3.2) which uses an integral formulation to derive the difference algorithms. Equation (6.7) is a consequence of this approach. It reflects numerically the fact that the triangle velocities must rotate and stretch as the grid rotates and stretches. The transformation $\mathfrak{R}$ is derived by considering the circulation about each vertex.

The boundaries of a vertex cell are defined by the side bisectors, as shown in Fig. 6-8. The vertex cell as defined in the figure has area

$$A_c = \sum_j \frac{1}{3} A_j, \tag{6.10}$$

where the sum extends over all adjacent triangles, exactly as in Eq. (6.3).

With this definition Eqs. (6.3) and (6.9) become

$$\nabla p_i = \frac{\frac{1}{3}\sum_j \nabla p_j A_j}{A_c}$$ (6.11)

and

$$\mathbf{v}_i = \frac{\frac{1}{3}\sum_j \mathbf{v}_j A_j}{A_c}.$$ (6.12)

Since the triangle velocities are constant over the triangle, the circulation taken about the boundary of the vertex cell is straightforward to calculate. The conservation of vorticity then takes the form of the operator $\mathfrak{R}$ in Eq. (6.7), which preserves the value of the circulation about each vertex while maintaining the proper divergence at each cell. This transformation ensures that the vorticity integral calculated about any interior vertex is invariant under the advancement of the grid. It is easy to show that the ∇p and gravity terms cannot alter the vorticity either since $\nabla \times \nabla p = 0$ identically and gravity is a constant. Only the $\nabla p_j / \rho_j$ term can change vorticity, exactly as dictated by the physics. Since the transformation $\mathfrak{R}$ is time reversible, Eqs. (6.4)–(6.8) are also. Hence the entire algorithm advances vertex positions and velocities reversibly while evolving the correct vorticity about every interior vertex.

This technique is unique for Lagrangian codes, which usually either ignore vorticity conservation completely or conserve vorticity through an iteration simultaneously with the pressure iteration. In this technique the vorticity is conserved exactly, regardless of whether the pressures have iterated to their final values.

The pressures p_i^n in Eq. (6.8) are derived from the condition that the new velocities $\mathbf{v}_j^n$ should be divergence-free at the new time step, satisfying Eq. (2.2). The pressure Poisson equation is derived from Eq. (6.8) by setting $\nabla \cdot \mathbf{v}_j^n = 0$ to obtain a pressure p_i^n such that

$$\frac{\delta t}{2} \nabla \cdot \frac{1}{\rho_j} (\nabla p)_j^n = \nabla \cdot \mathbf{v}_j^{1/2}.$$ (6.13)

The right-hand side of Eq. (6.13) is the numerical analog of the $\nabla \cdot (\mathbf{v} \cdot \nabla \mathbf{v}) \delta t$ term which arises when the divergence of Eq. (2.2) is taken. Both terms in Eq. (6.13) are simple to evaluate since the divergence is taken over triangle-centered quantities. The parts are the "surfaces" bounding the vertex cell of Fig. 6-8, where the normal is directed outward from the vertex. Two features of the Poisson equation (6.13) that results are noteworthy. First, it is derived from $\nabla^2 \phi = \nabla \cdot \nabla \phi$, as in the continuum case. Second, the left-hand side results in the more familiar second-order-accurate templates (such as the five-point formula) for the Laplacians found in connection with homogeneous fluids and regular mesh geometries.

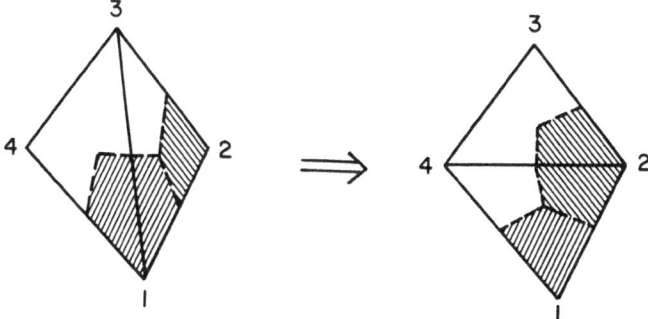

Fig. 6-9 Reconnection alters the vertex cells for each of the quadrilateral vertices.

The derivation of the reconnection and vertex addition and deletion algorithms is likewise accomplished through the control volume approach and
the use of triangle velocities. The accuracy of a computer code which uses a
reconnecting grid is determined by two aspects of the algorithms: how
accurately postreconnection physical variables are chosen, and when the
reconnections occur. As mentioned above, reconnection offers the possibility
of much less diffusion since fluid passes through only one grid line during
reconnections, whereas four grid lines are involved in every vertex rezone
movement. Every mesh line can be viewed as one diagonal of the quadrilateral
formed by the triangles to either side. A reconnection merely chooses the
other diagonal. During a reconnection the smallest definable cell is the quadrilateral, not the two triangles, and it is necessary to ensure that quadrilateral
properties are unchanged during a reconnection. That is, the quadrilateral is
a control volume over which certain physical variables are conserved. As
shown in Fig. 6-9, a reconnection alters the vertex cells for each of the four
quadrilateral vertices.

To conserve divergence and vorticity, the portions of the integrals $\int \nabla \cdot \mathbf{v} dA$
and $\int \nabla \times \mathbf{v} \cdot d\mathbf{A}$ about each vertex which lie within the quadrilateral must
be the same before and after the reconnection. Since there are four velocity
components to be specified after the reconnection, it would seem possible to
keep either the divergence or the vorticity about each vertex constant, but
not both. However, the divergence and vorticity equations are not all independent, so that it is possible in fact to choose a unique set of triangle velocities
which exactly conserve both vorticity and divergence about each vertex. This
same solution also conserves the quadrilateral velocity and has the added
feature of time reversibility. Repeatedly reconnecting a line yields the identical
grid and physical variables on both vertices and triangles. This is extremely
desirable since the basic finite-difference equations are also time reversible,
as are the physical equations themselves.

The question of when to reconnect remains unanswered. As mentioned

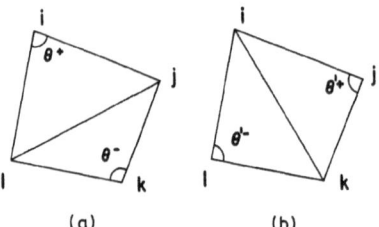

Fig. 6-10 Criterion for reconnection. If θ^+ and θ^- sum to more than 180° (a), the analogous angles θ'^+ and θ'^- formed after reconnection sum to less than 180° (b).

above, the accuracy of a general triangular mesh is diminished by narrow triangles. With reconnections, accuracy can be recovered by ensuring that small angles are preferentially eliminated. There are many ways of quantifying such an algorithm. The one we have chosen arises naturally from the pressure Poisson equation. Since the equation is solved by iteration, it is desirable that the convergence of the iteration be as rapid as possible. Mathematically, convergence is assured if the equation exhibits diagonal dominance. For a general triangular mesh, diagonal dominance is obtained only if all coefficients

$$a = \tfrac{1}{2}(\cot \theta^- + \cot \theta^+) \qquad (6.14)$$

are positive, where θ^+ and θ^- are defined in Fig. 6-10a. For positive area triangles θ^+ and θ^- are both between 0° and 180°, so that each term is negative only when $\theta^+ + \theta^- > 180°$, since

$$a = \frac{\sin(\theta^+ + \theta^-)}{2 \sin \theta^+ \sin \theta^-}. \qquad (6.15)$$

The reconnection algorithm is defined so as to exactly preserve diagonal dominance. If $\theta^+ + \theta^-$ is greater than 180°, the grid line is reconnected as shown in Fig. 6-10b. The new angles θ'^+ and θ'^- must sum to less than 180° since the sum of the interior quadrilateral angles is

$$\theta^+ + \theta^- + 0'^+ + 0'^- - 360°. \qquad (6.16)$$

Therefore, the reconnection algorithm is unique as well as straightforward. It also preferentially eliminates small angles for triangles since the diagonal is chosen which divides the largest opposing angles.

Because the reconnection algorithms are specified to ensure diagonal dominance and eliminate small-angle triangles, the second-order accuracy may be expected to be preserved. As yet no test has been made of the accuracy of the reconnection algorithms for the complete range of gridding situations. This is mainly because reconnections cannot themselves ensure second-order accuracy since flows near stagnation points must force narrow triangles in any case. That is, vertex addition and deletion must be employed at the same time.

The derivation of the remaining grid restructuring algorithms proceed in

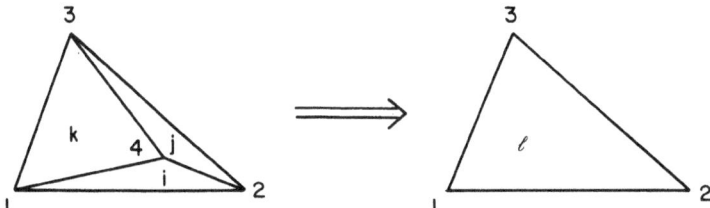

Fig. 6-11 Deletion of a vertex. After vertex 4 is isolated within a triangle by reconnections it can be eliminated. Deleting the vertex, three lines and the three triangles defines the larger triangle.

exactly the same manner as above. For vertex addition, satisfying the conservation integrals is particularly simple. A vertex added at the centroid of a triangle subdivides that triangle into three smaller triangles. If the new triangle velocities are the same as the velocity of the subdivided triangle, all conservation laws are trivially satisfied. Since the reconnection algorithm is also conservative, subsequent reconnections to other vertices ensure that the only effect of the addition is an increase in resolution.

The case is not as obvious for vertex deletion. Reconnections can be used to reduce the number of lines meeting at a given vertex to three. Elimination of these lines and the vertex leaves the surrounding triangle (Fig. 6-11). This new triangle is given a velocity which is the area-weighted sum of the old velocities,

$$A_l V_l = A_i \mathbf{v}_i + A_j \mathbf{v}_j + A_k \mathbf{v}_k, \tag{6.17}$$

where l labels quantities associated with the large triangle and i, j, k refer to the small ones. Such a substitution also conserves vorticity exactly and effects a redistribution in accordance with area coordinates. Figure 6-11 illustrates the triangles before and after vertex removal. If ζ_4 is the vorticity about vertex 4 before removal, then the vorticity about each of the other three vertices is increased by an amount ζ',

$$\zeta_1' = \frac{A_j \zeta_4}{A_l},$$

$$\zeta_2' = \frac{A_k \zeta_4}{A_l}, \tag{6.18}$$

$$\zeta_3' = \frac{A_i \zeta_4}{A_l},$$

where $\zeta_1' + \zeta_2' + \zeta_3' = \zeta_4$ since $A_i + A_j + A_k \equiv A_l$. Therefore, total vorticity is conserved and redistributed in a reasonable and natural manner. Since the behavior of the divergence is governed by a similar set of equations and conservation of momentum is ensured by Eq. (6.17), the conservation of the flow variables is guaranteed and the primary effect of deletion is a loss in resolution.

6.4 Examples

A finite-difference code SPLISH which embodies the algorithms discussed above has been tested on a series of physical problems. These problems were designed specifically to exercise particular algorithms as they were added and to test, as well, the overall accuracy of the code against well-known analytic and experimental solutions. For example, the reconnection algorithms have been tested extensively through calculations involving the Kelvin–Helmholtz instability. In the linear, nonlinear, and turbulent regimes, the algorithms have been shown to provide accurate calculations by comparisons with both theory and experiment (Fritts, 1976a; Fritts and Boris, 1979). The pressure generated by these algorithms, as well as the individual paths of vertices, have been compared with theory and experiment for progressive waves (Miner, *et al.*, 1979; Fritts, Miner, and Griffin, 1980).

An even more difficult test of a Lagrangian code has been carried out with an unstable density gradient. The Rayleigh–Taylor calculation for two fluids of density ratio 2:1 is patterned on the situation found in a simple laboratory demonstration and displays all of the linear and nonlinear features observed experimentally (Fig. 6-12).

An initial grid of triangles was established with a small sinusoidal perturbation of the interface at $t = 0$. The initial velocities were all taken to be zero. The flow field is followed as the instability enters its nonlinear phase. The vertex locations are plotted at successive time steps to give streaks whose lengths are proportional to the local fluid velocity. Shear flow is established as the interface steepens, giving rise to vortices due to the onset of the Kelvin–Helmholtz instability (Chandrasekhar, 1961). The two fluids are observed to mix in the vortices, entraining enough lighter fluid to form "bubbles," which are about to be completely enveloped by the heavier fluid.

Near the end of the calculation most of the grid carried by the lighter fluid has flowed out from below the downward-jetting heavier fluid. Similarly, vertices in the heavier fluid have been deleted above the upwelling lighter fluid. A reasonable resolution has been maintained by adding other vertices, particularly along the interface. Originally the interface was resolved with 10 vertices, but by the end of the run 46 were required. The heavier fluid has not touched the bottom, but rests upon extremely thinned and skewed triangles in the light fluid. The presence of these near-zero-area triangles would terminate the calculation due to violation of the Courant condition. At this stage a routine is invoked which locally eliminates the thinned layer and allows the dense fluid to contact the bottom. Similar problems must also be dealt with

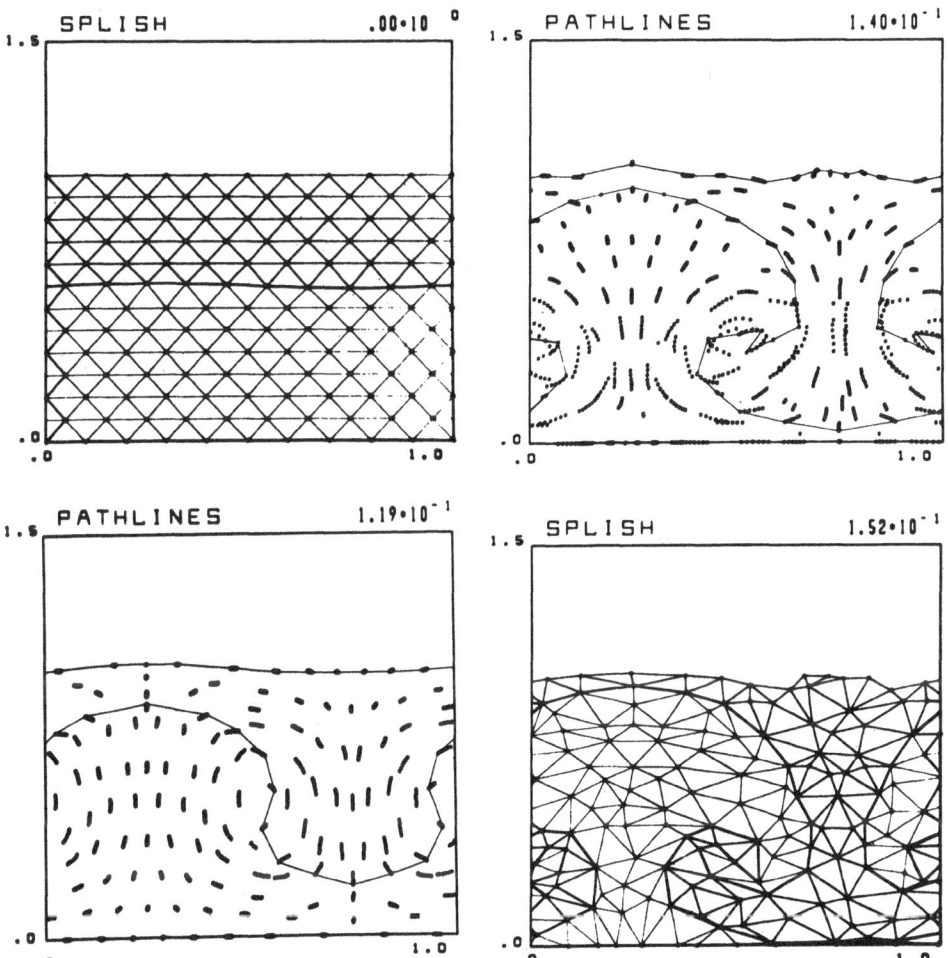

Fig. 6-12 Development of the Rayleigh–Taylor instability. Note the deformation of the interface occurring between the $t = 0.0$ and $t = 0.152$ frames, which show locations of the triangles. At $t = 0.119$ and $t = 0.140$ flow lines are marked to show the onset of shear flow and formation of Kelvin–Helmholtz-like vortices.

at the free surface near the upwelling light fluid and in the formation of the cavitation "bubbles" at the center of the vortices.

This approach may not be the only method of dealing with severe grid distortions (see, e.g., Chan, 1974), but it is the most general, most physical, and the least dissipative technique employed so far. Fritts (1976, 1976a) and Fritts and Boris (1977) have recently employed Lagrangian triangular grid techniques in studying nonlinear aspects of free-surface waves, including strongly sheared flows. Ten to 15 full cycles of waves can be integrated reversibly without significant deterioration of the solution In the presence of strong shear the reconnection procedure is reversible (i.e., nondissipative). Figure 6-13 shows a plot of the wave natural period versus grid size for six different resolution calculations of the same physical problem. The second-order accuracy over long times is demonstrated by the parabolic form of the curve. Going to a smaller step size gives improved accuracy, as expected. An added advantage accrues from using triangular cells. The nonlinear mesh-separation instabilities which plague low-dissipation rectangular-cell techniques seem to be absent when triangular cells are used.

The drawbacks to the triangular grid approach are the complexity of the programming required and the fact that the operations which are required seem to hinge strongly on linked lists, random access, and sequential processing. These problems become even more pronounced when one attempts to construct a three-dimensional hydrocode based on the use of tetrahedrons instead of triangles. The very structure of these codes would therefore seem to preclude any vectorization to accommodate the new pipeline and vector processors. But this is decidedly not the case.

The calculations performed within the general connectivity codes can be divided into two classes; calculations associated with the hydrodynamics equations and those involved with the grid manipulation. The bulk of the operations necessary to advance the physical variables can be vectorized with little effort. The difficulty associated with vectorizing the remaining portions of the hydrodynamics calculations can be isolated to two particular operations —storing and fetching data from memory in a scalar fashion. Experience has shown that these two operations can be optimized to run at speeds roughly seven times faster than purely scalar codes. Use of these two specialized routines to create input and output arrays for further vectorized calculations virtually completes the vectorization of a code. The only remaining operation needed is a vectorizable IF test. Provided a particular machine's hardware can support a vectorized IF test which stores indices of the expressions failing (or satisfying) the IF, this final obstacle to nearly complete vectorization can be removed.

Other trade-offs are necessary. For example, it is impossible to use successive overrelaxation (see Section 7.1) for vectorized calculations. However, a Jacobi iterative procedure (Varga, 1962) can be substituted. Although the Jacobi method is inherently slower to converge in terms of the number of iterations, the vastly improved speed of each iteration when vectorized can more than compensate, and its convergence is faster in terms of CPU time.

The second class of calculations which must be adapted to vector machines are the grid-restructuring routines. These operations are also inherently scalar. One grid reconnection must be completed before another is performed since both may involve the same original triangle. However, for most calculations only about 1% of the grid is altered for any time step, whereas all the grid

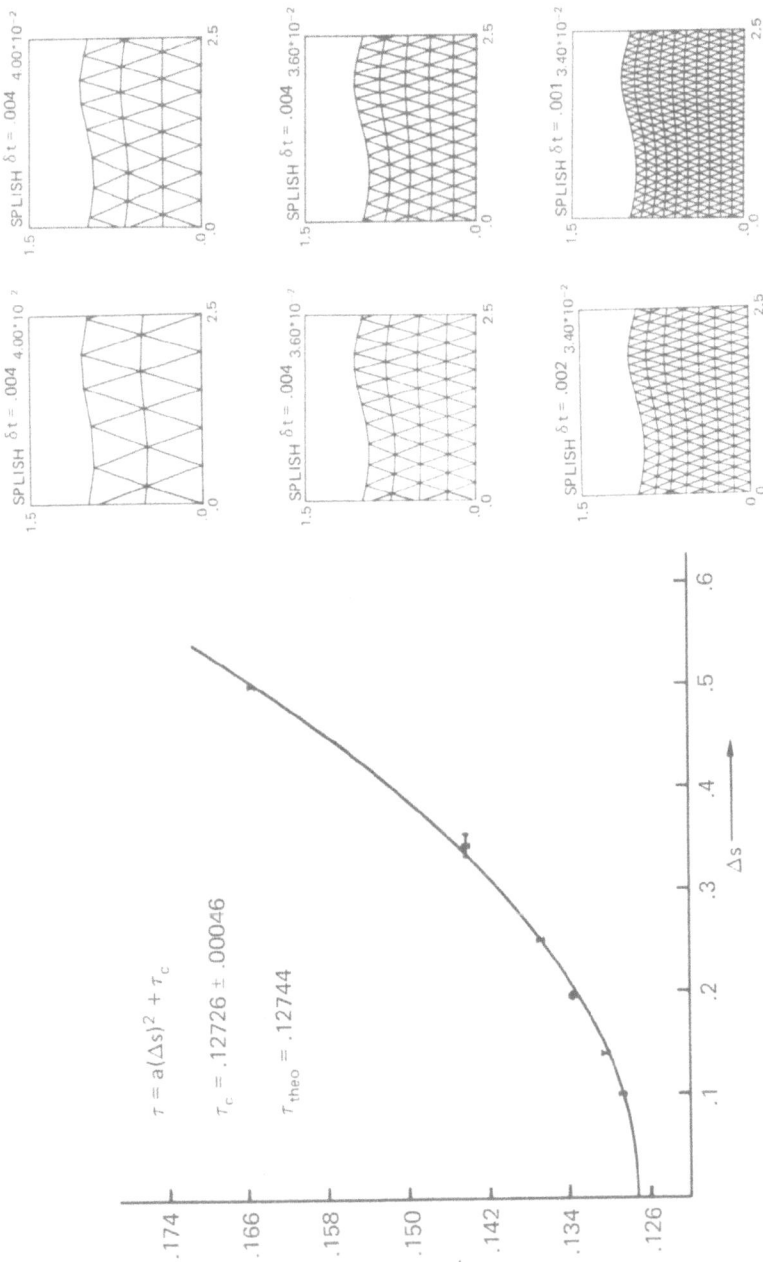

Fig. 6-13 Wave period versus normalized numerical grid size for six calculations of a finite-depth free-surface wave with different resolutions.

must be tested for potential alterations. These tests can be vectorized since no grid alteration results from the tests. A list of necessary alterations is then passed to the actual restructuring routines, which perform the necessarily scalar operations only on very small portions of the grid.

These techniques, coupled with the use of efficient input/output routines, have shown that general connectivity codes can be run quite efficiently on vector machines. Some of the vectorization comes easily, but specialized scalar routines and vectorized tests must be written. The greatest obstacle remaining is in finding more efficient algorithms for iterative relaxation of Poisson's equation. Although little work has been done using irregular grids, the outlook is promising with the recent introduction of techniques which are directly applicable, such as incomplete Cholesky solvers and the multigrid approach.

CHAPTER 7

Solution of Elliptic Equations

Solution of the Poisson equation (2.8) is necessary whenever we employ the vorticity–stream function formulation of the incompressible fluid equations. A Poisson equation (2.10) is also encountered when we seek the solution for the pressure in the primitive equation formulation. Generalizations of the Poisson equation arise in other physical problems with similar features, e.g., motion of plasma in the ionosphere (Scannapieco *et al.*, 1976; cf. Figs. 3-8 through 3-13). All of these are elliptic equations. In most calculations the boundary conditions take the form of Neumann, Dirichlet, or "radiation" conditions; in a few cases of interest, absorbing boundaries which neither transmit nor reflect waves are employed.

7.1 Survey of Standard Techniques

The techniques used for solving elliptic equations may be divided into "direct" and "iterative," or into "transform" and "finite-difference" types. Since transforms may be regarded merely as maximal-order finite-difference procedures, the former distinction is more fundamental for computational purposes. The present section is devoted to a discussion of the standard techniques from this point of view. Following Hockney (1970), we restrict our attention to the simple Poisson equation. Some of the methods discussed generalize straightforwardly to other types of elliptic equations.

Vector computers place a new set of constraints on the choice of numerical algorithms for computational problems. Factors such as the length of data vectors and the time which intervenes between obtaining a result and the subsequent use of that result as an operand in another computation must be considered in optimizing numerical algorithms. Sparse banded matrix equations, like those which arise from the finite-difference solution of elliptic equations on a two- or three-dimensional grid, are examples of the type of problem which can be treated by vectorizable matrix solvers.

Direct methods: The conceptually simplest direct method involves solving (2.8) via a double fast Fourier transform (Boris and Roberts, 1969). In **k**

space we have for a two-dimensional periodic system

$$\psi_{\mathbf{k}} = -k^{-2}\zeta_k. \tag{7.1}$$

In a finite-difference problem the proper representation of this for the usual five-point difference scheme is

$$\psi_{\mathbf{k}} = -\left[\frac{2}{\delta x^2}\,(1 - \cos k_x \delta x) + \frac{2}{\delta y^2}\,(1 - \cos k_y \delta y)\right]^{-1}\zeta_{\mathbf{k}}. \tag{7.2}$$

Inversion of the transform then completes the solution for ψ. The method generalizes readily to three dimensions [where (7.1) is replaced by three equations for the components of $\mathbf{A_k}$] and to simple nonperiodic boundaries through the addition of an appropriately chosen solution of Laplace's equation. The number of operations required on an $N_1 \times N_2$ mesh is $\sim 5\text{–}10$ times $N_1 N_2 [\log_2(N_1 N_2) - 3]$, the constant factor depending on the boundary conditions.

The Double Cyclic Reduction (DCR) method starts with the simple five-point two-dimensional finite-difference scheme, which may be written for a uniform mesh in the form

$$\mathbf{\Psi}_{j+1} - \mathbb{C}^{(1)} \cdot \mathbf{\Psi}_j + \mathbf{\Psi}_{j-1} = -\mathbf{S}_j^{(1)}, \tag{7.3}$$

where $\mathbf{\Psi}_j$ and $\mathbf{S}_j$ are column vectors and $\mathbb{C}^{(1)}$ is the tridiagonal matrix with coefficients $\{-1,\ 4,\ -1\}$. The first stage of odd/even reduction converts (7.3) into

$$\mathbf{\Psi}_{j+2} - \mathbb{C}^{(2)} \cdot \mathbf{\Psi}_{j-2} = -\mathbf{S}_j^{(2)}, \tag{7.4}$$

where

$$\mathbb{C}^{(2)} = \mathbb{C}^{(1)} \cdot \mathbb{C}^{(1)} - 2\mathfrak{I} \tag{7.5}$$

($\mathfrak{I}$ is the identity matrix) and

$$\mathbf{S}_j^{(2)} = \mathbf{S}_{j+1}^{(1)} + \mathbb{C}^{(1)} \cdot \mathbf{S}_j^{(1)} + \mathbf{S}_{j-1}^{(1)}. \tag{7.6}$$

The reduction process is repeated until the Lth level is reached ($N = 2^L + 1$), when only a single equation remains, which for Dirichlet boundary conditions relates the central vector to two known vectors of boundary values:

$$\mathbf{\Psi}_{M+1} = [\mathbb{C}^{(L)}]^{-1} \cdot [\mathbf{\Psi}_N + \mathbf{S}_{M+1}^{(L)} + \mathbf{\Psi}_1], \tag{7.7}$$

where $M = \frac{1}{2}(N - 1) = 2^{L-1}$. We now iterate (7.7) to find the column vector halfway between 1 and $M + 1$ and that halfway between $M + 1$ and N, then the column halfway between consecutive pairs of those columns, and so on. The expansion procedure at level l ($1 \le l < L$) can be stated for $j - 1$ a multiple of 2^{l-1} as

$$\mathbf{\Psi}_j = [\mathbb{C}^{(l)}]^{-1} \cdot [\mathbf{\Psi}_{j+m} + \mathbf{S}_j^{(l)} + \mathbf{\Psi}_{j-m}], \tag{7.8}$$

with $m = 2^{l-1}$ and all quantities on the right previously determined.

The matrices $\mathbb{C}^{(l)}$ have $2l + 1$ nonzero diagonals. They can be factored, however, as follows:

$$\mathbb{C}^{(2)} = [\mathbb{C}^{(1)}]^2 - 2\mathfrak{I} = [\mathbb{C}^{(1)} + \sqrt{2}\mathfrak{I}] \cdot [\mathbb{C}^{(1)} - \sqrt{2}\mathfrak{I}], \tag{7.9}$$

$$\mathbb{C}^{(3)} = [\mathbb{C}^{(2)} + \sqrt{2}\mathfrak{I}] \cdot [\mathbb{C}^{(2)} - \sqrt{2}\mathfrak{I}]$$

$$= [\mathbb{C}^{(1)} + (2 + \sqrt{2})^{1/2}\mathfrak{I}] \cdot [\mathbb{C}^{(1)} + (2 - \sqrt{2})^{1/2}\mathfrak{I}] \tag{7.10}$$

$$\cdot [\mathbb{C}^{(1)} - (2 - \sqrt{2})^{1/2}\mathfrak{I}] \cdot [\mathbb{C}^{(1)} - (2 + \sqrt{2})^{1/2}\mathfrak{I}],$$

and so on. It turns out that this factorization is impractical on 32-bit word machines for meshes larger than 128×128 because of overflow problems (the central coefficient of $\mathbb{C}^{(l)}$ is $\sim 4^m$), and other means of evaluating are preferable (Buneman, 1969).

All of the divisions by $\mathbb{C}^{(l)}$ in the series of equations (7.8) represent inversion of a tridiagonal matrix. Buneman's (1969) method involves tridiagonal inversion during both reduction and expansion which accounts for the name (DCR). Tridiagonal inversion can be performed by a variety of techniques, including odd/even reduction. It turns out that this is a particularly good tridiagonal inversion technique for use on the TI/ASC because parts of it vectorize readily. The total operation count for Buneman's (1969) symmetric form of DCR is approximately $6N_1 N_2 (\log_2 N_1 + 2)$, a modest improvement over double FFT.

A simple form of the Fourier Analysis–Cyclic Reduction (FACR) method begins with FFT in one direction, followed by odd/even (cyclic) reduction along each line in the transverse direction. The procedure for the reduction and expansion exactly parallels that just described. Fourier synthesis is then performed on the result to obtain ψ.

Hockney's (1965) version of FACR proceeds as follows. One stage of odd/even reduction is performed as before. Then a real FFT is performed on the even lines which result from the first step. The resulting equations form a tridiagonal system for each harmonic k. These are then solved by recursive application of cyclic reduction. (Gauss elimination is just as good on scalar computers, but not on pipeline or vector machines.) The Fourier synthesis is carried out, and finally the solutions on the odd lines are found from the original five-point formula by tridiagonal inversion.

The total number of operations is roughly $2.5N_1 N_2 [\log_2 N_1 + 2.5]$, again depending on the boundaries, details of the FFT, etc. The method can be generalized, however, to include l levels of odd/even reduction before Fourier analysis is performed (Hockney, 1970). When this is done, the total number of operations is $N_1 N_2 [3 + 4.5l + 2^{-l}(5 \log_2 N_1 - 4)]$. This has a shallow minimum at $l \gtrsim 2$ (depending on N_1, N_2) and shows that slight improvement over the preceding operation count is possible. Thus, for large arrays FACR is the fastest of these methods. All three require only one-dimensional scratch storage, as they overwrite the source array with the potential solution.

These methods are not restricted to inversion of two-dimensional Laplacians. Wilhelmson and Ericksen (1977) discuss direct solution of the Poisson equation (2.10) in three dimensions (i.e., when the Laplacian is three dimensional) by various combinations of cyclic reduction and Fourier analysis, presenting timings and operation counts.

Another method for inversion of sparse banded (not necessarily penta-diagonal) $N \times N$ matrices is Crout (1941) reduction. Given a general sparse equation of the form

$$\mathfrak{M} \cdot \mathbf{z} = \mathbf{s}, \tag{7.11}$$

write M as a product of upper and lower triangular matrices:

$$\mathfrak{M} = \mathfrak{L} \cdot \mathfrak{U}. \tag{7.12}$$

Then solve

$$\mathfrak{L} \cdot \mathbf{y} = \mathbf{s}, \tag{7.13}$$

followed by

$$\mathfrak{U} \cdot \mathbf{z} = \mathbf{y}. \tag{7.14}$$

In the coded algorithm, the vectorized decomposition operates on the rectangle whose vertices are on the main diagonal and the outermost bands. This rectangle moves down the main diagonal one step at a time. Thus, the kernel of the decomposition is a triply nested DO–loop on a vector dot product, the fastest vector instruction on the ASC. The solutions are simply row-by-row dot products.

During exection the dot product in the decomposition dominates the calculation. It requires approximately $4N^3$ operations (additions plus multiplications times the length of the main diagonal times the bandwidth). For $N \sim 40$, the computing time is about 10% more than $2N^3$ times the theoretical vector speed. The decomposed matrix occupies $2N^3$ storage locations and dominates the memory requirement.

Iterative methods: Iterative methods are easy to code and (depending on the problem) can accelerate convergence by careful choice of the starting guess. The best known of these is successive overrelaxation (SOR) by points. For example, if we write the five-point approximation to the Poisson equation in the form

$$R_{ij} = \psi_{i+1,j} + \psi_{i-1,j} + \psi_{i,j+1} + \psi_{i,j-1} - 4\psi_{ij} - \zeta_{ij} = 0. \tag{7.15}$$

the iterative scheme is defined by

$$\psi_{ij}^{(new)} = \psi_{ij}^{(old)} + \tfrac{1}{4}\omega R_{ij}, \tag{7.16}$$

where ω is an adjustable number. For this scheme, the rate of convergence is highly sensitive to the order in which the corrections are carried out. If they are done a line at a time the procedure can be vectorized. Typically 5–10 iterations can be performed during the time required for a direct solution.

The worst-case rate of convergence (based on the poorest possible initial guess) can be improved by varying ω suitably. One technique, the Cyclic Chebyshev method, uses the following prescription to vary ω every half iteration:

$$\omega^o = 1, \tag{7.17}$$

$$\omega^{1/2} = (1 - \tfrac{1}{2}\mu^2)^{+1}, \tag{7.18}$$

$$\omega^{n+1/2} = (1 - \tfrac{1}{4}\mu^2\omega^n)^{-1}, \tag{7.19}$$

where μ is fixed. However, worst-case analyses have little to do with physically well-motivated calculations.

Quite different is the Incomplete Cholesky-Conjugate Gradient (ICCG) method (Kershaw, 1978). For a general sparse matrix operator $\mathfrak{M}$, we solve (7.11) for $\mathbf{z}$ as follows. First find triangular matrices $\mathfrak{L}$ and $\mathfrak{U}$ which approximately satisfy (7.12). We require that $L_{ij} = 0$ at every location (i, j) where $M_{ij} = 0$. It is necessary that the diagonal elements be nonvanishing, even if this introduces an error in the representation. Let

$$\mathfrak{R}_1 = \mathfrak{U}^T \cdot \mathfrak{U} \tag{7.20}$$

and

$$\mathfrak{R}_2 = \mathfrak{L} \cdot \mathfrak{L}^T, \tag{7.21}$$

where the superscript T denotes the transverse of a matrix. Define an initial guess for the solution, for example, by

$$\mathbf{z} = (\mathfrak{L} \cdot \mathfrak{U})^{-1} \cdot \mathbf{s}. \tag{7.22}$$

Set

$$\mathbf{r} = \mathbf{s} - \mathfrak{M} \cdot \mathbf{z} \tag{7.23}$$

and

$$\mathbf{p} = \tilde{\mathfrak{M}}^T \cdot \mathbf{r}, \tag{7.24}$$

where

$$\tilde{\mathfrak{M}}^T = \mathfrak{R}_1^{-1} \cdot \mathfrak{M}^T \cdot \mathfrak{R}_2. \tag{7.25}$$

Then let

$$\gamma = \mathbf{r} \cdot \mathfrak{R}_1^{-1} \cdot \mathbf{r} \tag{7.26}$$

and

$$\alpha = \frac{\gamma}{\mathbf{p} \cdot \mathfrak{R}_2 \cdot \mathbf{p}}. \tag{7.27}$$

Now iterate:

$$\mathbf{z}' = \mathbf{z} + \alpha\mathbf{p}, \tag{7.28}$$

$$\mathbf{r}' = \mathbf{r} - \alpha\mathfrak{M}\cdot\mathbf{p}, \tag{7.29}$$

$$\beta = \gamma^{-1}\mathbf{r}'\cdot\mathfrak{R}_1^{-1}\cdot\mathbf{r}', \tag{7.30}$$

$$\mathbf{p}' = \mathfrak{M}^T\cdot\mathbf{r}' + \beta\mathbf{p}, \tag{7.31}$$

until $\|\mathfrak{M}\cdot\mathbf{z} - \mathbf{s}\|$ is sufficiently small. In the algorithm all operations are expressed in terms of vector instructions except the calculations of the two inverses, $\mathfrak{R}_1^{-1}$ and $\mathfrak{R}_2^{-1}$, which require solutions of an upper triangular system of equations and a lower triangular system of equations. These operations are totally scalar since $\mathfrak{L}$ is sparse by construction.

During execution, the computing time for the vector operations is small compared with the two upper-triangular and two lower-triangular solutions. Thus, the required time is approximately $8N^2 N_d N_i$ times the scalar operation speed, where N_d is the number of diagonals (above or below) and N_i is the number of iterations. In addition to the diagonals, the right-hand side, and the solution vector, the algorithm requires about nine more vectors. The data storage is thus of order $(N_d + 9)N^2$.

A number of iterative techniques for solving elliptic equations are related to those used to advance diffusion equations in time. Thus, if an initial guess for ψ is advanced in time by means of a finite-difference approximation to

$$\frac{\partial\psi}{\partial t} = \nabla^2\psi - \zeta \tag{7.32}$$

with the appropriate boundary conditions until a stationary state is reached, the resulting function is the solution of (2.8). Of the various methods for solving parabolic equations, the one most often used is Alternating Direction Implicit (ADI). For (7.32) it can be represented as

$$\psi_{ij}^{1/2} = \psi_{i,j}^o + \frac{\Delta t}{2}\left[\frac{\psi_{i+1,j}^{1/2} - 2\psi_{i,j}^{1/2} + \psi_{i-1,j}^{1/2}}{\Delta x^2}\right.$$
$$\left. + \frac{\psi_{i,j+1}^o - 2\psi_{i,j}^o + \psi_{i,j-1}^o}{\Delta y^2} - \xi_{i,j}^o\right], \tag{7.33}$$

$$\psi_{i,j}^n = \psi_{i,j}^{1/2} + \frac{\Delta t}{2}\left[\frac{\psi_{i+1,j}^o - 2\psi_{i,j}^o + \psi_{i-1,j}^o}{\Delta x^2}\right.$$
$$\left. + \frac{\psi_{i,j+1}^n - 2\psi_{i,j}^n + \psi_{i,j}^n - 1}{\Delta y^2} - \zeta_{i,j}^o\right]. \tag{7.34}$$

On the alternate time steps, the implicit differencing is performed in the y-direction instead of the x-direction, whence the name.

The optimum choice of Δt gives slightly faster convergence than SOR, although with more operations. By varying the interval Δt appropriately from one step to the next, convergence can be improved. No simple rules are available for this. The number of iterations required for convergence of ADI varies only weakly with the size of the mesh.

Other iterative techniques are discussed by Roache (1972). Their usefulness, as well as that of many of the variants of SOR, ICCG, and ADI, depends in large measure on the details of the problem being solved.

7.2 A New Direct Solver: The Stabilized Error Vector Propagation Technique (SEVP)

Elliptic partial differential equations may be separable or nonseparable and their solutions are subject to a variety of boundary conditions. If we write the elliptic finite-difference form of the differential equation in the form (7.11), where $\mathbf{s}$ is the forcing function and $\mathbf{z}$ is the required solution, then $\mathfrak{M}$ can be diagonally dominant, weakly diagonally dominant, or even non-diagonally dominant. Even though a number of iterative and direct solvers are available in the literature for solving elliptic equations, no one method is efficient for all types of equations. The iterative methods, such as SOR and ADI, and hybrid methods, such as optimized block implicit relaxation (Dietrich, McDonald, and Warn-Varnas, 1975), are very efficient for diagonally dominant equations. However, when diagonal dominance is weak and residual error much in excess of roundoff cannot be tolerated, convergence rates for these methods may be unattractively slow. In addition, many iterative methods will diverge when applied to nondiagonally dominant equations.

Direct solvers based on FACR, DCR, and the method of Rosmond and Faulkner (1976) are very powerful but are restricted to separable equations. For nonseparable equations the direct solvers of Lindzen and Kuo (1969) and Crout (1941; see Section 7.1) are available but require large amounts of computer memory and thus are limited to small arrays. The error vector propagation (EVP) method (Roache, 1971, 1977; Hirota, Tokioka, and Nishiguchi, 1970) is applicable to any type of elliptic equation and requires an order of magnitude less memory than the Lindzen–Kuo method. However, as shown by McAvaney and Leslie (1972), EVP is unstable when the number of grid points is large. Here we describe a method due to Madala (1978), called stabilized error vector propagation (SEVP), which is stable for any number of grid points and retains most of the advantages of EVP.

A general two-dimensional second-order elliptic equation in finite-difference form can be written as

$$a_{i,j}\psi_{i,j-1} + b_{i,j}^1\psi_{i-1,j} + b_{i,j}^2\psi_{i,j} + b_{i,j}^3\psi_{i+1,j} + c_{i,j}\psi_{i,j+1} = S_{i,j}, \qquad (7.35)$$

where the coefficients a, b^1, b^2, b^3, and c may be functions of both i and j,

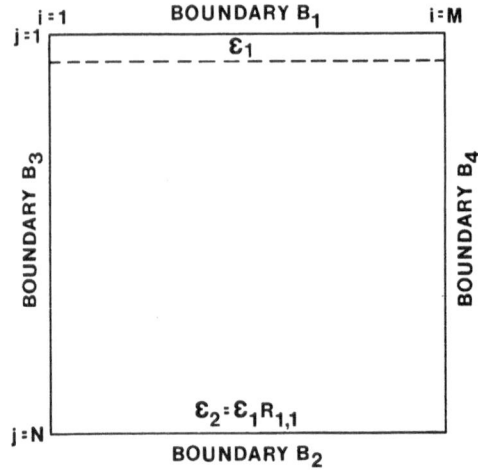

Fig. 7-1 The region over which the elliptic equation is solved by the EVP method. B_1, B_2, B_3, and B_4 are the boundaries.

and S is the forcing function. We now review the EVP procedure described by Roache (1971, 1977) for solving (7.35) over a rectangular region (shown in Fig. 7-1) using Dirichlet boundary conditions. By rearranging the terms, we can rewrite (7.35) as

$$\psi_{i,j+1} = \frac{S_{i,j} - a_{i,j}\psi_{i,j-1} - b_{i,j}^1\psi_{i-1,j} - b_{i,j}^2\psi_{i,j} - b_{i,j}^3\psi_{i+1,j}}{c_{i,j}}. \tag{7.36}$$

It is clear from (7.36) that if we know the solution on any two consecutive rows, then we can march in j to obtain the required solution. Since the solution is known only on the boundaries B_1, B_2, B_3, and B_4, the EVP method requires an initial guess of the solution on the row interior to B_1 to march forward in j. Based on this initial guess a particular solution is obtained which satisfies (7.36) and all the boundary conditions except along B_2. If we assume that the particular solution $\psi_{i,j}^P$ deviates from the real solution by $\psi_{i,j}^H$, then

$$\psi_{i,j} = \psi_{i,j}^P + \psi_{i,j}^H. \tag{7.37}$$

Substituting (7.37) into (7.36) and noting that the particular solution satisifies (7.36) and the boundary conditions at B_1, B_3, and B_4, we obtain the following homogeneous equation:

$$\psi_{i,j+1}^H = -\frac{a_{i,j}\psi_{i,j-1}^H + b_{i,j}^1\psi_{i-1,j}^H + b_{i,j}^2\psi_{i,j}^H + b_{i,j}^3\psi_{i+1,j}^H}{c_{i,j}}, \tag{7.38}$$

with zero boundary conditions along B_1, B_3, and B_4. Along B_2 we find from (7.38)

$$\psi_{i,N}^H = \psi_{i,N} - \psi_{i,N}^P. \tag{7.39}$$

We can relate the homogeneous solution on any two rows by using $M - 2$ independent vectors, where M represents the total number of grid points in i, including boundary points. Since the solution along B_2 [given by (7.39)] is known, we will relate it to the solution on the row just interior to the boundary B_1. In other words, if the vectors ε_1 and ε_2 (Fig. 7-1) represent the homogeneous solution on the second and last rows, respectively, then we can find an influence matrix $\mathfrak{R}_{1,1}$ such that

$$\varepsilon_2 = \varepsilon_1 \cdot \mathfrak{R}_{1,1}. \tag{7.40}$$

To obtain $\mathfrak{R}_{1,1}$, we start with the kth unit vector along the second row and march (7.38) in j with zero boundary conditions along B_1, B_3. The kth row elements of the $\mathfrak{R}_{1,1}$ matrix are the elements of ψ^H along the boundary B_2. By varying the values of k from 2 to $M - 1$ we complete $\mathfrak{R}_{1,1}$.

From (7.40) and the error vector [given by (7.39)] we compute the homogeneous solution on the second row. Using these values on the second row and zero boundary conditions on B_1, B_3, and B_4 we march (7.38) in j to obtain the homogeneous solution throughout the region. Then by superposition of the particular and homogeneous solutions we obtain the complete solution. Due to roundoff errors in computing the inverse of the matrix $\mathfrak{R}_{1,1}$, this solution does not satisfy the boundary condition along B_2 exactly. The accuracy of the solution is improved by recomputing the homogeneous part of the solution a few times. (Thus, this method, though direct, usually involves iteration.) Normally, about six iterations are required for 20 grid points in the marching direction to obtain an error of 10^{-14} on a computer with a 56-bit word precision. The number of iterations required increases as the number of grid points in the marching direction is increased up to about 25; beyond 25 the method becomes unstable.

The influence matrix $\mathfrak{R}_{1,1}$ depends only on the coefficients of the elliptic equation and therefore can be computed without knowledge of the forcing function and the boundary conditions. This step is called the preprocessor. If we want to solve (7.35) repeatedly with the same coefficients but with different forcing functions and boundary conditions, the influence matrix needs to be computed only once and stored in the memory for further use.

Although the EVP method is very efficient and requires a very small amount of computer memory, it is unstable for a large number of grid points in the marching direction (McAvaney and Leslie, 1972). For computers with 24-bit precision, the limit is about 12 grid points, which may be increased by going to higher precision or by using the two-directional marching method described by Hirota, Tokioka, and Nishiguchi (1970). Even with 56-bit precision and two-way marching, the limit is extended only to about 40 grid points.

In the SEVP method, the integration region is divided into blocks, each of which is small enough for the EVP method to be stable. Further, any two consecutive blocks have two rows in common. The division of the integration

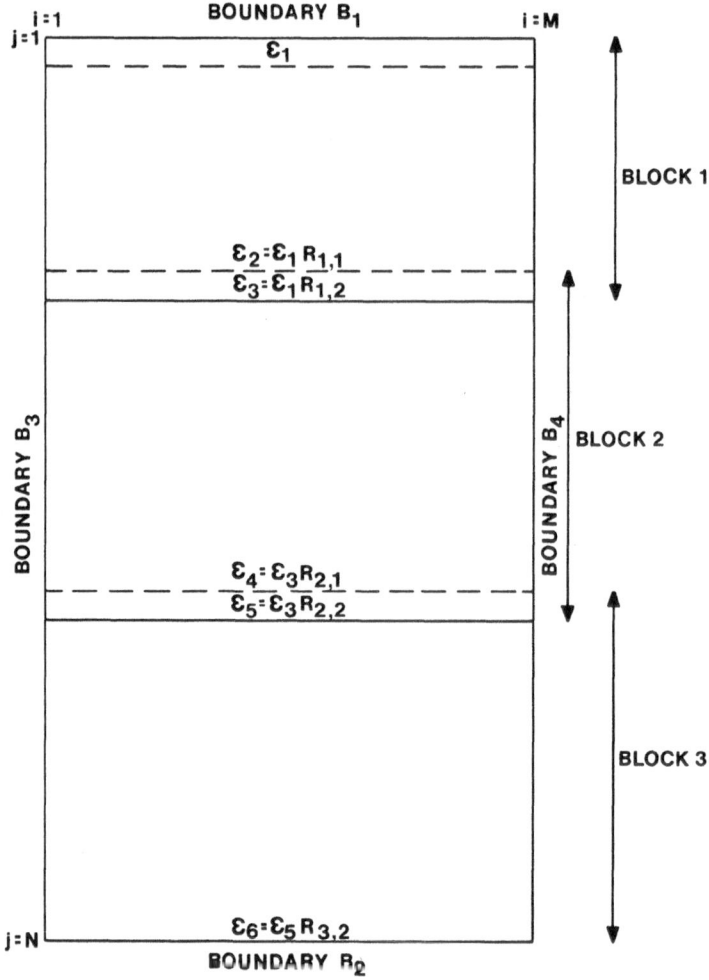

Fig. 7-2 A three-block division of the region for the SEVP method.

region into three blocks is shown in Fig. 7-2. As with the EVP method, SEVP is divided into two steps: the preprocessor step and the solution step. In the preprocessor step, $2N_b - 1$ influence matrices are computed, where N_b is the total number of blocks. N_b of the influence matrices relate the values of the homogeneous solution on the second and last rows of each block, while the remaining $N_b - 1$ matrices relate the homogeneous solution on the second rows of consecutive blocks.

The first $M - 2$ unit vectors on the second row of the first block are used to march in the j-direction, using (7.38) and zero boundary conditions on

B_1, B_3, and B_4 to obtain influence matrices for the last two rows of the block. If ε_1, ε_2, and ε_3 (shown in Fig. 7-2) represent the homogeneous solution on the second and last two rows of the first block, then

$$\varepsilon_2 = \varepsilon_1 \cdot \mathfrak{R}_{1,1} \tag{7.41}$$

and

$$\varepsilon_3 = \varepsilon_1 \cdot \mathfrak{R}_{1,2}, \tag{7.42}$$

where $\mathfrak{R}_{1,1}$ and $\mathfrak{R}_{1,2}$ are the influence matrices for the last two rows of the block.

Combining (7.41) and (7.42) to eliminate ε_1 we obtain

$$\varepsilon_3 = \varepsilon_2 \cdot \mathfrak{R}_{1,1}^{-1} \cdot \mathfrak{R}_{1,2} = \varepsilon_2 \cdot \mathfrak{Q}_1. \tag{7.43}$$

The matrices $\mathfrak{R}_{1,1}$ and $\mathfrak{Q}_1$ relate homogeneous solutions on the second and last rows of the block and on the last two rows of the block, respectively.

To obtain the influence matrices for the second block, we start with the unit vectors on the second row of the second block, and compute the corresponding values on the first row of the block from (7.43) to get the equivalent of a boundary condition for this block. The homogeneous solution on the first row corresponding to the kth unit vector on the second is the kth row of the matrix $\mathfrak{Q}_1$. Using $M - 2$ unit vectors on the second row with corresponding vectors on the first row and zero boundary values on B_3 and B_4, we advance (7.38) in j to obtain influence matrices for the last two rows of the second block. If the vectors ε_3, ε_4, and ε_5 represent a homogeneous solution on the second and last two rows of the block, respectively, then

$$\varepsilon_4 = \varepsilon_3 \cdot \mathfrak{R}_{2,1} \tag{7.44}$$

and

$$\varepsilon_5 = \varepsilon_3 \cdot \mathfrak{R}_{2,2}, \tag{7.45}$$

where $\mathfrak{R}_{2,1}$ and $\mathfrak{R}_{2,2}$ are the influence matrices for the last two rows of the block. Eliminating ε_3 from (7.44) and (7.45) we obtain

$$\varepsilon_4 = \varepsilon_5 \cdot \mathfrak{Q}_2, \tag{7.46}$$

where

$$\mathfrak{Q}_2 = \mathfrak{R}_{2,2}^{-1} \cdot \mathfrak{R}_{2,1}. \tag{7.47}$$

Repeating the procedure described above for the third block (last block in Fig. 7-2) we will obtain a matrix, $\mathfrak{R}_{3,2}$, relating the homogeneous solution on the second and last rows of the block. If ε_5 and ε_6 represent the homogeneous solution on these rows, then

$$\varepsilon_6 = \varepsilon_5 \cdot \mathfrak{R}_{3,2}. \tag{7.48}$$

As in the EVP method, all the influence matrices depend only on the coefficients of (7.35).

We obtain the required solution by one forward and one backward sweep of the blocks. During the forward sweep, an approximate solution is obtained which satisfies the equation and the boundary conditions everywhere except on boundary B_2. In the backward sweep, this solution is corrected to obtain the exact solution.

The forward sweep is started with the second row on the first block with an initial guess for the elements of that row and the given boundary values along the first row. This procedure is exactly the same as that for EVP. When the solution has been marched to the end of the first block, arbitrary values are assigned along the block boundary to obtain the correction vector ε_3. The second sweep forward generates a homogeneous solution for the first block. Iterations are applied, if necessary, to reduce roundoff error.

For the second block we take the arbitrary solution assigned along the first block boundary (second row, second block) and the solution along the first row of the second block and sweep forward again. An arbitrary solution is again imposed along the last row of the second block to obtain the correction vector ε_5. On the second (or homogeneous) sweep of the second block we use the correction vectors ε_2 and ε_3 to form the homogeneous solution. We compute ε_3 from ε_5 by (7.45) and obtain ε_2 from ε_3 using (7.43). The procedure continues until all blocks are swept out. For the last block, we use the boundary B_2 instead of a guess value to find ε_6.

The particular solution we have obtained on the forward sweep satisfies (7.35) and the given boundary conditions everywhere except on the block boundaries. More importantly, the last block contains the exact solution since the computation of the correction vector ε_6 is based upon the known boundary conditions along B_2.

The backward sweep now corrects the errors introduced in the particular solution by guessing the solutions along the block boundaries. Using (7.45) and (7.46) we calculate ε_4 and ε_5 for the homogeneous sweep of the last block and obtain the total solution for the last block, which has two rows in common with the second block. To obtain the homogeneous solution for the second block we need to find the difference ε_5' between the particular solution on the last row of the block and the exact solution. One method of obtaining ε_5' is to store the particular solution from the forward sweep and subtract it from the exact solution. However, it is much easier to recompute the particular solution by backing up three rows into the second block and repeating a small segment of the forward sweep. After ε_5' is obtained we sweep the second block to obtain the homogeneous solution, which is then added to the particular solution. This procedure is repeated until we finish the blocks. Error reduction iterations are applied at this step also.

The stabilization of EVP by this method is achieved by the introduction of the artificial boundaries. Propagation of error is very severe in EVP, and when the error exceeds the accuracy of the computer it cannot be corrected by addition of a homogeneous solution. The introduction of artificial bound-

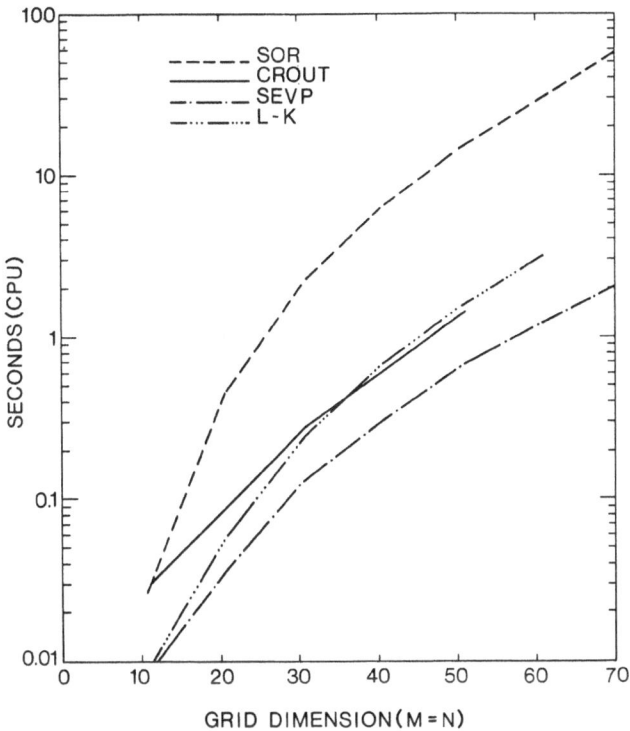

Fig. 7-3 A comparison of the computing time requirements for the SEVP, Lindzen–Kuo, Crout, and SOR methods.

aries before the error becomes too large limits the error in each block. Since the influence matrices for small blocks have small error levels, the artificial solution generated by the introduction of these boundaries can be easily corrected to obtain an extremely accurate final solution.

As a test problem for SEVP, Poisson's equation is solved over a square region with zero homogeneous boundary conditions. The test problem is also solved with the Lindzen–Kuo (Lindzen and Kuo, 1969), Crout (1941), and SOR methods. The first two are direct solvers. The relaxation by SOR is stopped when the error (normalized with the forcing function) reaches 10^{-4} A comparison of the computing time taken by these methods and SEVP is given in Fig. 7-3. When the grid dimension exceeds 60 for Lindzen–Kuo (L–K) and 50 for Crout methods the computer memory is exhausted. It is clear from the figure that the SEVP method is more efficient than the other three methods. For $M = N = 40$, the SEVP method is about two times faster than L–K and Crout methods and 20 times faster than SOR. The computation times given in Fig. 7-3 were obtained with double precision

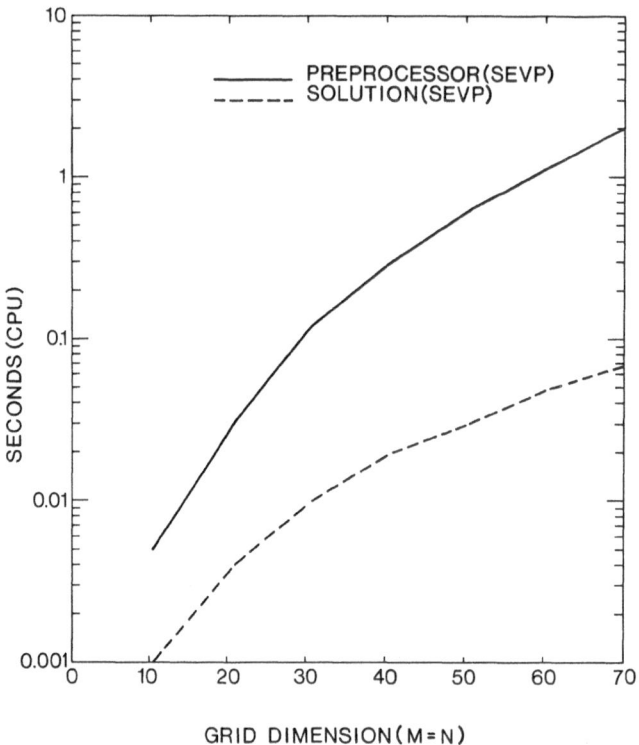

GRID DIMENSION (M = N)

Fig. 7-4 Computer time taken by the preprocessor and solution steps of the SEVP method.

(56-bit precision) operations on the Texas Instruments ASC, using the version of the algorithm implemented in the subroutines listed as Appendix D.

Figure 7-4 shows the computer time required by the preprocessor step and the solution step of SEVP on the same vector computer. The ratio of computing time required by the preprocessor step to that required by the solution step is about 3 for $M = N = 10$ and increases with increase in grid dimension. The L–K method can also be separated into a preprocessor step (computing α matrices) and a solution step. For the test problem single precision on a 32-bit-word computer is adequate for the L–K method.

Figure 7-5 gives a comparison of the computer time taken by the single precision Lindzen–Kuo and double precision SEVP methods to solve the test problem. The top two curves are the total computer time (preprocessor and solution steps) required, while the lower two curves are for the solution steps only. It is clear from the figure that the SEVP method is faster than the Lindzen–Kuo method for both the preprocessor and solution steps. ICCG (not shown) performed slightly worse than Crout because of relatively incomplete vectorization.

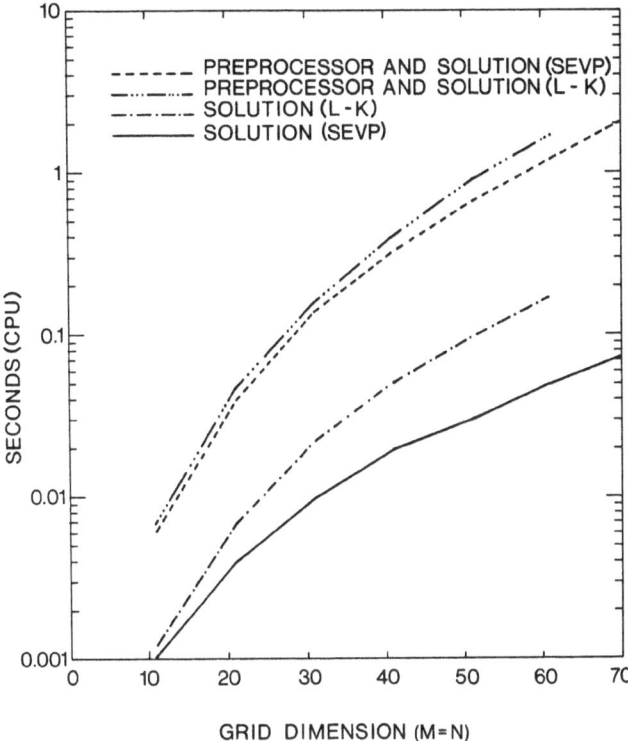

Fig. 7-5 A comparison of the computer time requirements of single precision Lindzen–Kuo and double precision SEVP methods using a vector computer with 32-bit-word length. The top curves give the total time taken by the preprocessor and the solution steps together. The lower two curves give the time taken by the solution steps.

Table 7-1 gives the auxiliary memory requirements for the three direct methods for a 32-bit-word computer using double precision. It is clear from the table that the SEVP method's auxiliary memory requirement is an order of magnitude smaller than that required by the other two direct methods.

If N_K represents the maximum number of points in the marching direction of the Kth block, then

$$N_K \leq PA, \tag{7.49}$$

where P is a constant which depends only on the computer precision and A represents the minimum value of $|c_{i,j}/b_{i,j}^2|$. It is clear from this equation that we can reduce the auxiliary memory required by the SEVP method by choosing the marching direction in such a way that $A \geq 1$. It can also be shown that the method requires less computing time to solve the elliptic equation in the case when $A \geq 1$ than in cases when $A < 1$. Therefore, the

Table 7-1 A comparison of the auxiliary memory requirements (in thousands of words) for SEVP, Lindzen–Kuo and Crout methods on 32-bit-word computer using double precision. The numbers in parentheses exceed the total central memory available on the NRL ASC.

Grid size	Crout	SEVP	Linzen–Kuo
10	4.4	0.2	2.8
20	33.2	2.4	17.6
30	111.6	5.4	57.6
40	262.4	16.0	134.4
50	510.0*	25.0	260.0
60	(878.4)	50.4	446.4
70	(1391.6)	88.2	(705.6)
80	(2073.6)	115.1	(1049.6)
90	(2978.4)	178.2	(1490.4)
100	(4040.0)	220.0	(2040.0)

* Normalized error is more than 10^{-4}.

computing speed of the method increases with A. On the other hand, the Lindzen–Kuo and Crout methods deteriorate with the increasing A and become unstable for $A \geq 100$.

The memory requirements for SEVP can be reduced greatly by making simple modifications. These modifications closely follow those developed by Schoeberl (1980) for the L–K method. There it was shown that the influence matrices following any two consecutive rows can be derived from the relationship of the two rows. Instead of solving the influence matrices relating every two blocks, we save one matrix for every few blocks and recompute the rest whenever they are needed. This can be done by grouping the blocks. We divide these N_b blocks into L groups, each similar to Fig. 7-2. All these groups need not have the same number of blocks. However, we will show later that the memory requirements are minimum when all the groups are of the same size.

At any time, the influence matrices required are those between the groups and all the influence matrices in any one group. If N_b^L represents the maximum of the number of blocks in the groups, then the number of words of memory M_{tot} required for SEVP is

$$M_{tot} = (L - 1 + 2N_b^L - 1)(N - 2)^2, \tag{7.50}$$

where L represents the total number of groups. M_{tot} is a function of L and N_b^L. If all the groups are of the same size, then Eq. (7.50) reduces to

$$M_{tot} = L - 2 + 2\frac{N_b}{L}(N - 2)^2. \tag{7.51}$$

M_{tot} is minimum for $L = (2N_b)^{1/2}$, which gives

$$M_{tot}^{min} = 2[(2N_b)^{1/2} - 1](N - 2)^2. \tag{7.52}$$

This procedure for solving elliptic equations with Dirichlet boundary conditions can easily be extended to other boundary conditions (Neumann, periodic, or mixed). It can also be used to solve elliptic equations over a region with irregular boundaries.

Another kind of generalization is also possible. Two-dimensional elliptic difference equations with nine-point stencils (over three or five consecutive rows and columns) can be solved by using the EVP and SEVP methods with a few minor modifications. For an elliptic difference equation with a nine-point stencil over three consecutive rows and columns, a general equation can be written

$$
\begin{aligned}
c_{i,j}^1 \psi_{i-1,j+1} &+ c_{i,j}^2 \psi_{i,j+1} + c_{i,j}^3 \psi_{i+1,j+1} \\
&= F_{i,j} - a_{i,j}^1 \psi_{i-1,j-1} - a_{i,j}^2 \psi_{i,j-1} - a_{i,j}^3 \psi_{i+1,j-1} \\
&\quad - b_{i,j}^1 \psi_{i-1,j} - b_{i,j}^2 \psi_{i,j} - b_{i,j}^3 \psi_{i+1,j}.
\end{aligned}
\tag{7.53}
$$

Earlier, to obtain the homogeneous and particular solutions, we used the marching procedure in the j-direction, computing the solution explicitly at every point in the $(j + 1)$th row in terms of the jth and $(j - 1)$th rows. Since (7.53) now couples the solution at three consecutive points on the $(j + 1)$th row, we instead compute the solution at all points on the $(j + 1)$th row simultaneously with a tridiagonal solver. With this minor modification, the EVP and SEVP methods described can be used to solve (7.53). The use of the tridiagonal solver increases the number of operations by $8(M - 2)^2(N - 1)$ for the preprocessor step and by $8(M - 2)^2(2 + \alpha)$ for the solution step, where α is the number of iterations.

A general elliptic difference equation using nine-point stencil over five consecutive rows and columns can be written

$$
\begin{aligned}
c_{i,j}^2 \psi_{i,j+2} = F_{i,j} &- a_{i,j}^1 \psi_{i,j-1} - a_{i,j}^2 \psi_{i,j-2} - b_{i,j}^1 \psi_{i-2,j} - b_{i,j}^2 \psi_{i-1,j} \\
&- b_{i,j}^3 \psi_{i,j} - b_{i,j}^4 \psi_{i+1,j} - b_{i,j}^5 \psi_{i+2,j} - c_{i,j}^1 \psi_{i,j+1}.
\end{aligned}
\tag{7.54}
$$

The solution of this fourth-order finite-difference equation requires four boundary conditions in each direction, which may be specified as the solution on the two outermost rows and columns of the region. As before we obtain the required solution by superposing a particular and a homogeneous solutions. The particular solution is obtained by marching in the jth direction using the given solution (boundary condition) on the first two rows and a guess solution on the next two rows. The result satisfies (7.54) with the given forcing function and the boundary condition everywhere except on the boundary representing the last two rows. Since the homogeneous solution is the deviation of the particular solution from the required solution, it will satisfy (7.54) with zero forcing function everywhere and homogeneous boundary conditions on all the boundaries except on the last two rows. On these two rows the boundary conditions are obtained by subtracting the particular solution from given boundary conditions. Following the procedure described earlier using the homogeneous equation, we compute an influence matrix relating a vector representing the homogeneous solution vectors on the third and fourth rows with that on the last two rows. Since each of these vectors has $2(M - 4)$ grid points, we use $2(M - 4)$ unit vectors to obtaining the influence matrix. Using the influence matrix and the boundary conditions for the homogeneous solution on the last two rows, we can obtain the error solution on the third and fourth rows. With these values for the error solution on the third and fourth rows and zero boundary conditions on the first two rows, first two columns, and last two columns, we can obtain the solution for the homogeneous equation everywhere in the interior by marching in the jth direction.

The modifications described above to solve Eq. (7.54) by the EVP method can be easily incorporated in the SEVP procedure. As before, the region is divided into a number of blocks in such a way that each block is stable for the EVP method. The first and last two rows and columns of each block are the boundaries. Thus, any two consecutive blocks now have four rows in common. The preprocessor step consists of computing influence matrices for each block that relate the homogeneous solution on the third and fourth rows to the last two rows. We also compute an influence matrix for every two consecutive blocks to relate the homogeneous solution on the first two common rows with the homogeneous solution on the next two common rows. During the forward sweep of the solution step we obtain an approximate solution, given guess values at the block boundaries where the solution is unknown. During the backward sweep these boundary conditions are connected to obtain the required solution. The procedure is similar to the one described earlier.

The size of each influence matrix described above is approximately four times the size of the corresponding influence matrix associated with Eq. (7.36). Thus, the EVP and SEVP methods require four times more auxiliary memory to solve a fourth-order finite-difference equation than a second-order equation. It can also be shown that the preprocessor step takes approximately eight times more computing time and the solution step takes approximately four times more computing time.

7.3 Application of Chebychev Iteration to Non-Self-Adjoint Equations

For some problems requiring the solution of an elliptic equation, obtaining a solution to roundoff accuracy may not be required. If a certain level of residual error can be tolerated, and a good trial solution is available (as in the case of many time integration problems), an iterative solution of sufficient accuracy may sometimes be obtainable in a fraction of the time required for a complete solution. This generally requires that the equation be well conditioned in some sense and that the residual error not have a cumulative effect on parameters of interest. We will not attempt to give criteria here for determining when iteration is preferable to direct solution, but present a method that has been used successfully in that context (McDonald *et al.*, 1978).

The computational mathematics literature is well stocked with iterative methods for solution of self-adjoint linear operator equations. However, many fewer approaches are offered for non-self-adjoint problems. One occasionally encounters the suggestion that the equation

$$\mathcal{L}\psi(r) = S(r), \tag{7.55}$$

with $\mathcal{L}$ the linear operator and S the known driving term be made self-adjoint by an extra application of the adjoint of $\mathcal{L}$:

$$\mathcal{L}^+\mathcal{L}\psi = \mathcal{L}^+S. \tag{7.56}$$

There may be cases in which this approach has merit. However, it has two immediate drawbacks: (1) one has to work with a higher-order equation, and (2) the ratio of maximum to minimum eigenvalue amplitude, an indicator of the amount of numerical work required, is squared. On the other hand, there are situations in which it may be desirable to transform a self-adjoint problem into a non-self-adjoint one. For example, in a temperature or salinity diffusion problem, $\mathcal{L}$ contains the operator $\nabla \cdot K\nabla$, where K is the diffusivity. If K varies greatly in magnitude, the eigenvalue span of $\mathcal{L}$ will be increased accordingly, making the numerical solution more difficult. A possible remedy is to divide through by K, so that $\mathcal{L}$ contains the non-self-adjoint operator $\nabla^2 + \nabla \ln K \cdot \nabla$.

With some adaptation, ICCG may hold promise for non-self-adjoint problems. But the factorization stage is awkward for vector computers. As indicated earlier, factorization of a matrix into upper and lower triangular matrices requires many scalar operations. Thus, we wish to describe an alternative method developed by McDonald (1980) which allows complete vectorization (parallel processing) for all interior mesh points of a multidimensional grid. This method is a generalization of the Chebychev semi-iterative method (Varga, 1962). Unlike ADI, its convergence rate holds regardless of whether the operator can be split into one-dimensional operators which commute with one another.

We seek an iterative solution to Eq. (7.55). We assume $\mathcal{L}$ has complex eigenvalues $\lambda_r + i\lambda_i$, with all λ_r being of the same sign. We also assume all λ fall within an ellipse in the complex plane (see Fig. 7-6) whose major axis coincides with the real axis. The intersections of the ellipse with the real axis are $b - a$ and $b + a$ with

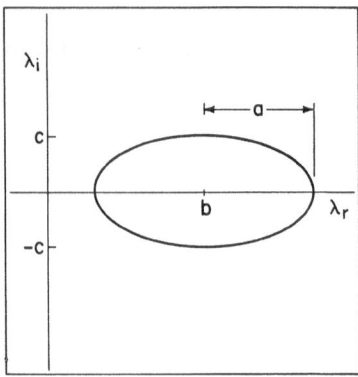

Fig. 7-6 Ellipse containing complex eigenvalues.

$$\frac{b}{a} > 1. \tag{7.57}$$

The semiminor axis of the ellipse is c. The equation of the ellipse is thus

$$\left(\frac{\lambda_r - b}{a}\right)^2 + \left(\frac{\lambda_i}{c}\right)^2 \leq 1. \tag{7.58}$$

The method to be developed is best implemented when

$$\left|\frac{c}{a}\right| < 1. \tag{7.59}$$

An alternative formulation will be given for cases in which this inequality does not hold.

The "best" values of a, b, and c are dependent on the nature of $\mathscr{L}$. For some cases, the error convergence rate may depend strongly on these values. We shall later give a physically motivated example in which "best" values can be estimated straightforwardly. However, if one does not have a priori estimates, values can be obtained numerically by systematic means (Manteuffel, 1977).

We assume that (7.55) possesses an exact solution Ψ. At the end of n iterations, we will have an approximate solution Ψ^n, whose error is defined to be

$$\varepsilon^n = \psi^n - \Psi. \tag{7.60}$$

The iterative method is to be such that

$$\varepsilon^n = P_n(\mathscr{L})\varepsilon^0, \tag{7.61}$$

where P_n is a polynomial of degree n. Substitution of (7.60) into (7.61) gives

$$\phi^n = P_n(\mathscr{L})\psi^0 - [P_n(\mathscr{L}) - 1]\Psi. \tag{7.62}$$

We do not know ψ in advance, but we do know $\mathscr{L}^k\psi = \mathscr{L}^{k-1}S$ for $k > 0$. Thus from (7.62) we must require that the zero-degree term of $P_n(\mathscr{L})$ be 1:

$$P_n(0) = 1. \tag{7.63}$$

We wish to choose P_n such that its magnitude is as small as possible everywhere within the ellipse containing eigenvalues of $\mathscr{L}$.

The min-max problem: The problem of minimizing the maximum value of $|P_n(x)|$ subject to $P_n(0) = 1$ has a well-known solution when x is restricted to real values between $b - a$ and $b + a$ with $b/a > 1$. The standard argument (Manteuffel, 1977) points out that the desired P_n is such that all maxima of $|P_n|$ have the same value. One immediately determines that P_n is proportional to a Chebychev polynomial. The Chebychev polynomials T_n are such that

$$T_n(\cos \alpha) = \cos n\alpha, \tag{7.64}$$

and

$$T_n(\cosh \alpha) = \cosh n\alpha. \tag{7.65}$$

That $\cos n\alpha$ is a polynomial in $\cos \alpha$ results from the elementary identity

$$\cos(m + 1)\alpha = 2 \cos \alpha \cos m\alpha - \cos(m - 1)\alpha. \tag{7.66}$$

Equation (7.66) is also valid when cosh is substituted for cos. Thus, if $\cos m\alpha$ and $\cos(m - 1)\alpha$ are polynomials in $\cos \alpha$, then so is $\cos(m + 1)\alpha$. This is the case for $m = 0$ and 1, so it is also true for all $m > 0$. Equation (7.66) also gives the recursion formula for the Chebychev polynomials:

$$T_{m+1}(x) = 2xT_m(x) - T_{m-1}(x), \tag{7.67}$$

with

$$T_0(x) = 1 \tag{7.68}$$

and

$$T_1(x) = x. \tag{7.69}$$

Note that $T_n(x)$ is an even or odd function of x, as n is even or odd. From (7.64) and (7.65) we can see that $|T_n(x)|$ reaches a limiting value of unity, $n + 1$ times as x varies from -1 to 1. Thus the solution to the min–max problem for real x is

$$P_n(x) = T_n\left(\frac{x - b}{a}\right)\left[T_n\left(-\frac{b}{a}\right)\right]^{-1} \tag{7.70}$$

and

$$|P_n|_{\max} = \left| T_n\left(-\frac{b}{a}\right)\right|^{-1}$$
$$= \left[\cosh n \cosh^{-1}\left(\frac{b}{a}\right)\right]^{-1} < 1. \tag{7.71}$$

The Chebychev polynomials also possess optimal properties in the complex plane. With

$$z = \xi + i\eta,$$ (7.72)

we seek a polynomial $P_n(z)$ of degree n such that $P_n(0) = 1$ and such that the maximum $|P_n(z)|$ will be as small as possible within the ellipse

$$\left(\frac{\xi - b}{a}\right)^2 + \frac{\eta^2}{c^2} = 1.$$ (7.73)

For $c/a < 1$, the unique solution to this min–max problem is (Manteuffel, 1977)

$$P_n(z) = T_n\left(\frac{z - b}{\sqrt{a^2 - c^2}}\right) \left[T_n\left(\frac{-b}{\sqrt{a^2 - c^2}}\right)\right]^{-1}.$$ (7.74)

We shall demonstrate that for a given c/a less than or greater than unity, every maximum of $|P_n|$ in the ellipse (7.73) has the same value, and that this maximum is less than unity.

For $c/a < 1$, we first point out that all n roots of P_n are pure real. As a result, $|P_n(z)|$ increases monotonically away from the real axis; i.e., there are no local maxima. This means we need examine only the boundary of the ellipse (7.73) for maxima.

Express the argument of T_n in the numerator of (7.74) as

$$z' = \frac{z - b}{\sqrt{a^2 - c^2}}$$

$$= \cos\alpha \cosh\beta + i\sin\alpha \sinh\beta$$ (7.75)

$$= \cos(\alpha - i\beta),$$

where α and β are real. Proof that an arbitrary complex number can be expressed in this form with real α and β is straightforward.

Comparing (7.72) and (7.75) and eliminating α one can show that surfaces of constant β are ellipses in the complex plane:

$$\left(\frac{\xi - b}{\cosh\beta\sqrt{a^2 - c^2}}\right)^2 + \left(\frac{\eta}{\sinh\beta\sqrt{a^2 - c^2}}\right)^2 = 1.$$ (7.76)

This ellipse is identical to (7.73) when

$$\tanh\beta = \frac{c}{a},$$ (7.77)

so that

$$\cosh\beta = \left(1 - \frac{c^2}{a^2}\right)^{-1/2}.$$ (7.78)

Note from (7.75) that z' runs once around the ellipse as α varies from 0 to 2π. From (7.75) we find

$$T_n(z') = \cos(n\alpha - in\beta)$$
$$= \cos n\alpha \cosh n\beta + i \sin n\alpha \sinh n\beta,$$
(7.79)

so that

$$|T_n(z')|^2 = \cosh^2 n\beta - \sin^2 n\alpha.$$
(7.80)

This shows that on an ellipse of constant β, $|T_n(z')|$ reaches the same maximum value, namely, $\cosh n\beta$, $2n + 1$ times as α varies from 0 to 2π. The maximum $|P_n(z)|$ can now be obtained from (7.74):

$$|P_n(z)|_{max} = \frac{\cosh n \cosh^{-1}(1 - c^2/a^2)^{-1/2}}{\cosh n \cosh^{-1}(|b|/a)(1 - c^2/a^2)^{-1/2}} < 1.$$
(7.81)

We have used the even or odd symmetry of T_n to remove the minus sign from the argument in the denominator of (7.74). The inequality in (7.81) results from $|b|/a > 1$ and the monotonicity of the functions cosh and $\cosh^{-1}$. For $c/a > 1$, (7.74) becomes

$$P_n(z) = T_n\left(\frac{z - b}{i\sqrt{c^2 - a^2}}\right)\left[T_n\left(\frac{-b}{i\sqrt{c^2 - a^2}}\right)\right]^{-1}.$$
(7.82)

All n roots of P_n are now on a line parallel to the imaginary axis, and there are still no local maxima in $|P_n(z)|$. Again, we need examine only the perimeter of the ellipse for maxima. The argument in the numerator of (7.82) can be expressed as

$$z' = -\frac{i(z - b)}{\sqrt{c^2 - a^2}}$$
$$= \cos \alpha \cosh \beta + i \sin \alpha \sinh \beta$$
$$= \cos(\alpha - i\beta).$$
(7.83)

Comparing (7.72) and (7.83) and eliminating α we find

$$\left(\frac{\eta}{\cosh \beta \sqrt{c^2 - a^2}}\right)^2 + \left(\frac{\xi - b}{\sinh \beta \sqrt{c^2 - a^2}}\right)^2 = 1.$$
(7.84)

This is identical to (7.23) when

$$\tanh \beta = \frac{a}{c},$$
(7.85)

$$\cosh \beta = \left(1 - \frac{a^2}{c^2}\right)^{-1/2}.$$
(7.86)

Using (7.83), we find

$$|T_n(z')|^2 = \cosh^2 n\beta - \sin^2 n\alpha. \tag{7.87}$$

This gives $\cosh n\beta$ as the maximum amplitude of the numerator in (7.82). We can evaluate the denominator in (7.82) by setting $z = 0$, $\alpha = \tfrac{1}{2}\pi$, and $\sinh \beta = b/\sqrt{c^2 - a^2}$ in (7.83). For even n, (7.87) gives the amplitude of the denominator as $\cosh n \sinh^{-1} b/\sqrt{c^2 - a^2}$, while for odd n the result is $\sinh n \sinh^{-1} b/\sqrt{c^2 - a^2}$. For this reason we must restrict n to even values to guarantee convergence when $c/a > 1$. For even n, (7.81) gives

$$
\begin{aligned}
|P_n(z)|_{\max} &= \frac{\cosh n \cosh^{-1}(1 - a^2/c^2)^{-1/2}}{\cosh n \sinh^{-1} b/\sqrt{c^2 - a^2}} \\[2mm]
&= \frac{\cosh n \cosh^{-1}(1 - a^2/c^2)^{-1/2}}{\cosh n \cosh^{-1}(1 + (b^2 - a^2)/c^2)^{1/2}(1 - a^2/c^2)^{-1/2}} \tag{7.88} \\[2mm]
&< 1.
\end{aligned}
$$

To illustrate the necessity of taking n even, let us consider an example in which $b/c = 0.9$ and $a/c = 0.899$. For $n = 16$, (7.88) gives $|P_n|_{\max} = 0.984$. However, with $n = 17$, we must change the cosh function in the denominator of (7.88) to sinh, and we have $|P_n|_{\max} = 1.003$.

The algorithm: We shall take the polynomial of (7.74) to be the proper choice for use in constructing the approximate solution to (7.55) according to (7.62). For reasons of numerical stability we choose to construct $P_n(\mathscr{L})$ through the use of recursion formulas rather than by factorization. The only real drawback to this method is that one extra array of storage is required for retention of earlier iterates. We substitute (7.74) into (7.62) to obtain

$$\psi^{n+1} = T_{n+1}\left(\frac{\mathscr{L} - b}{\sqrt{a^2 - c^2}}\right)\left[T_{n+1}\left(\frac{-b}{\sqrt{a^2 - c^2}}\right)\right]^{-1}(\psi^o - \Psi) + \Psi. \tag{7.89}$$

In the numerator of this expression let us express T_{n+1} in terms of T_n and T_{n-1} according to the recursion formula (7.67)–(7.69), and then express T_n and T_{n-1} in terms of ψ^n and ψ^{n-1} according to (7.89). Then we find

$$\psi^{n+1} = \frac{2T_n(q)}{T_{n+1}(q)\sqrt{a^2 - c^2}}[(\mathscr{L} - b)\psi^n - S] - \frac{T_{n-1}(q)}{T_{n+1}(q)}\psi^{n-1}, \tag{7.90}$$

where

$$q = -\frac{b}{\sqrt{a^2 - c^2}}. \tag{7.91}$$

Notice that the coefficient of Ψ is zero. Recall that we imposed this condition at the outset in (7.63).

In order to use this method we must know the first two terms in the recursion.

We get these by direct construction of $P_n(\mathscr{L})$ and obtain

$\psi^0 = $ arbitrary trial solution,

$$\psi^1 = \psi^0 - \frac{1}{b}(\mathscr{L}\psi^0 - S). \tag{7.92}$$

If $c > a$, then q in (7.91) becomes imaginary. We can avoid complex arithmetic in calculating T_n in (7.87) by use of the Chebychev polynomials of imaginary argument:

$$T_n(ix) \equiv i^n \tau_n(x). \tag{7.93}$$

Substitution of (7.93) into (7.67)–(7.69) gives the recursion formula

$$\tau_{m+1}(x) = 2x\tau_m(x) + \tau_{m-1}(x), \tag{7.94}$$

where

$$\tau_0(x) = 1 \tag{7.95}$$

and

$$\tau_1(x) = x. \tag{7.96}$$

When $c > a$, Eq. (7.90) is replaced with

$$\psi^{n+1} = \frac{2\tau_n(q')}{\tau_{n+1}(q')\sqrt{c^2 - a^2}}[(\mathscr{L} - b)\psi^n - S] + \frac{\tau_{n-1}(q')}{\tau_{n+1}(q')}\psi^{n-1}, \tag{7.97}$$

where $q' = b/\sqrt{c^2 - a^2}$. In the limit $c \to a$, q and q' tend to infinity and both (7.90) and (7.97) give

$$\psi^{n+1} = \psi^n - \frac{1}{b}(\mathscr{L}\psi^n - S). \tag{7.98}$$

Error estimate: If ϕ is an eigenvector of $\mathscr{L}$ such that

$$\mathscr{L}\phi = \lambda\phi, \tag{7.99}$$

where λ is a complex constant, then

$$P_n(\mathscr{L})\phi = P_n(\lambda)\phi. \tag{7.100}$$

Let us imagine that the error ε^0 of (7.60) is expressed as a linear combination of the eigenvectors of $\mathscr{L}$. We have assumed all eigenvalues of to lie within an ellipse (7.73) in the complex plane. Then from (7.61) and (7.81) or (7.88), the coefficients of the eigenvector expansion of ε^n have each been decreased in amplitude by at least a factor

$$|P_n(\lambda)| \leq \frac{T_n(a/\sqrt{a^2 - c^2})}{T_n(b/\sqrt{a^2 - c^2})} \equiv E_n(a,b,c). \tag{7.101}$$

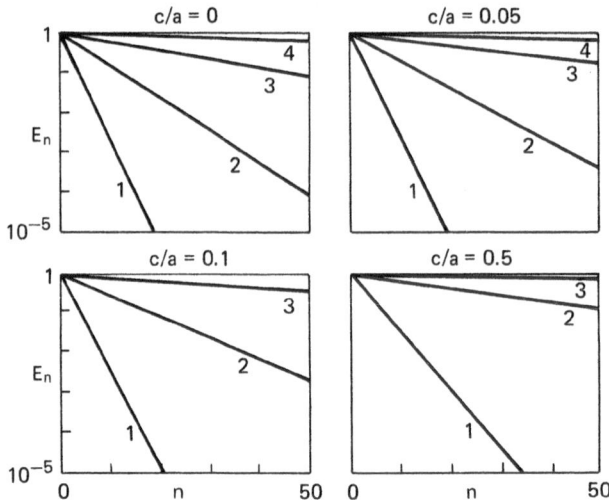

Fig. 7-7 Error estimate E_n versus n. Curves are labelled with $\log_{10} R$.

In Fig. 7-7 we plot this error limit as a function of n for various values of the ratio of maximum to minimum real eigenvalue,

$$R \equiv \frac{|b| + a}{|b| - a}, \tag{7.102}$$

and ellipse axis ratio c/a. One notices that the error limit decreases approximately exponentially with n.

Let us derive an approximate expression for (7.101) in the limit of large R. The results to be given here are valid for arbitrary $c/a \neq 1$. It is convenient to define

$$\begin{aligned} \delta &= \frac{|b|}{a} - 1 \\ &= \frac{2}{(R - 1)} \end{aligned} \tag{7.103}$$

and

$$\rho = \frac{c}{a}.$$

From (7.65) and the identity

$$\cosh^{-1} x = \log(x + \sqrt{x^2 - 1}), \tag{7.104}$$

we have from (7.81) or (7.88), as appropriate,

$$E_n = \frac{\cosh(n \log Q_1)}{\cosh(n \log Q_1 + n \log Q_2)} \tag{7.105}$$

$$= [\cosh(n \log Q_2) + \tanh(n \log Q_1) \sinh(n \log Q_2)]^{-1},$$

where

$$Q_1 = \frac{1 + \rho}{\sqrt{|1 - \rho^2|}}, \tag{7.106}$$

$$Q_2 = \frac{1 + \delta + \sqrt{2\delta + \delta^2 + \rho^2}}{1 + \rho}. \tag{7.107}$$

Since Q_1 and Q_2 are both greater than 1, (7.105) yields

$$2 \exp(-n \log Q_2) > E_n > \exp(-n \log Q_2). \tag{7.108}$$

Thus, the estimate

$$E_n \approx \sqrt{2} \left(\frac{1 + \delta + \sqrt{2\delta + \delta^2 + \rho^2}}{1 + \rho} \right)^{-n} \tag{7.109}$$

is always correct within a factor of $\sqrt{2}$. Defining the convergence rate to be

$$C = -\frac{\partial \log E_n}{\partial n}, \tag{7.110}$$

we can find simple expressions for C in the limit $\delta \ll 1$. For $\rho \ll \delta$, $C \approx \sqrt{2\delta}$. When $\rho \gg \delta$, then $C \approx \delta/\rho$. This shows that convergence slows down when the eigenvalues of the operator $\mathscr{L}$ have significant imaginary parts.

Application to a model problem: As a test of the method (7.90) and the error estimate (7.101) we solve numerically the two-dimensional diffusion-advection equation

$$\nabla^2 \psi + \mathbf{A}(x,y) \cdot \nabla \psi = S(x,y) \tag{7.111}$$

on a rectangular grid with constant mesh spacing subject to doubly periodic boundary conditions. Other boundary conditions may be treated by changing the value of a single parameter σ (to be defined below). Equations of this form arise in plasma physics when charge neutrality is imposed upon a plasma which has a nonisotropic electrical conductivity, with ψ being the electrostatic potential. For an application to the earth's ionosphere, see McDonald et al. (1975).

The mesh is assumed to be N_x by N_y interior mesh points. We choose the x-direction so that

$$N_x \geq N_y. \tag{7.112}$$

We also assume the mesh intervals to be

$$\delta x = \delta y = \text{const.} \tag{7.113}$$

We use redundant guard cells around the perimeter of the mesh for efficient handling of the boundary conditions. This results in a complete mesh of $(N_x + 2) \times (N_y + 2)$ points. We use the second-order five-point representation (7.35) in the form

$$
\begin{aligned}
\mathscr{L}\psi_{i,j} \equiv\ & \frac{\psi_{i+1,j} + \psi_{i-1,j} + \psi_{i,j+1} + \psi_{i,j-1} - 4\psi_{i,j}}{\delta x^2} \\
& + \frac{A_x^{i,j}(\psi_{i+1,j} - \psi_{i-1,j})}{2\delta x} \\
& + \frac{A_y^{i,j}(\psi_{i,j+1} - \psi_{i,j-1})}{2\delta x}
\end{aligned}
\tag{7.114}
$$

Here i and j are the x and y mesh point indices, and A_x and A_y are the x- and y-compoments of $\mathbf{A}$, respectively.

In order to estimate the eigenvalues of $\mathscr{L}$ it is necessary to consider $\mathbf{A}$ locally constant. Then the eigenfunctions are complex exponentials,

$$
\psi_n = \exp 2\pi i \left[\left(\frac{n_x i}{N_x} + \frac{n_y j}{N_y} \right) \right],
\tag{7.115}
$$

where

$$
n_x = 0, 1, 2, \ldots, N_x - 1,
\tag{7.116}
$$

and similarly for n_y. The eigenvalues are

$$
\begin{aligned}
\lambda_n =\ & \frac{2}{\delta x^2} \left(\cos \frac{2\pi n_x}{N_x} + \cos \frac{2\pi n_y}{N_y} - 2 \right) \\
& + i \left(\frac{A_x}{\delta x} \sin \frac{2\pi n_x}{N_x} + \frac{A_y}{\delta x} \sin \frac{2\pi n_y}{N_y} \right).
\end{aligned}
\tag{7.117}
$$

Let us now proceed to define ellipse parameters a, b, c so as to maximize the convergence rate (7.110). Any set of eigenvalues of bounded modulus can be enclosed in a sufficiently large ellipse. However, for the Chebychev iteration to have a positive convergence rate, the ellipse must exclude the origin. In the analysis to follow we are motivated by the need to produce an algorithm for solving the variable coefficient equation (7.111) at each time step. The coefficients change in time, so it does not seem appropriate to spend a great deal of effort in calculating precise iteration parameters for a particular $\mathbf{A}(x,y)$. Rather we will assume that an appropriate global measure of $|\mathbf{A}|$ is available, and from this construct parameters optimized for the most unfavorable orientation of $\mathbf{A}$. The resulting convergence rate is positive, but its proximity to the maximum convergence rate for specific distributions $\mathbf{A}(x,y)$ has not been investigated in a systematic way.

In order to find the tightest ellipse containing the complex eigenvalues for arbitrary orientation of $\mathbf{A}$, let us adopt the following abbreviated notation:

$$c_x = \cos\frac{2\pi n_x}{N_x},$$

$$s_x = \sin\frac{2\pi n_x}{N_x},$$

$$a_x = A_x \delta x, \qquad\qquad\qquad\qquad (7.118)$$

$$a_y = A_y \delta y,$$

$$\xi' = c_x + c_y,$$

$$\eta' = a_x s_x + a_y s_y.$$

From (7.117), $\lambda_n = (2/\delta x^2)(\xi' - 2) + (i/\delta x^2)\eta'$. To find the envelope of these eigenvalues we need only maximize $|\eta'|$ for fixed ξ'. The orientation of (a_x, a_y) must be allowed to be arbitrary (we are solving a variable coefficient equation and must allow for the "worst case"). Maximizing on the orientation of (a_x, a_y) and on c_x, we find the envelope

$$\lambda_e = \frac{2}{\delta x^2(\xi' - 2)} \pm \frac{i}{\delta x^2 \bar{a}(2 - \tfrac{1}{2}\xi'^2)^{1/2}}, \qquad (7.119)$$

where $\bar{a} = (a_x^2 + a_y^2)^{1/2}$. $\qquad\qquad\qquad\qquad (7.120)$

This is an ellipse with parameters $(a, b, c) = (4, -4, \bar{a}\sqrt{2})/\delta x^2$. It is the tightest ellipse containing all the eigenvalues for arbitrary N_x, N_y, a_x, and a_y subject to fixed $\bar{a}$. Unfortunately, it passes through the origin and thus results in a zero convergence rate. By modifying the ellipse, however, we can obtain a positive convergence rate.

Maximizing the convergence rate: We must exclude from the ellipse the "mean value" component $(n_x, n_y) = (0, 0)$, which has no effect upon (7.111). Let us then construct a family of ellipses passing through $\lambda_{0r} \pm i\lambda_{0i}$, the eigenvalue pair with real part nearest zero. Maximizing the convergence rate with respect to the two remaining free ellipse parameters amounts to maximizing Q_2 in (7.107). A significant simplification results if we approximate (7.107) as follows for $\delta \ll 1$:

$$Q_2 \approx 1 + \sqrt{2\delta + \rho^2} - \rho. \qquad\qquad\qquad (7.121)$$

Equation (7.121) is valid to lowest order if δ for all $\rho \geq 0$. The family of ellipses passing through $\lambda_{0r} \pm i\lambda_{0i}$ has

$$a = \left[(\lambda_{0r} - b)^2 + \left(\frac{\lambda_{0i}}{\rho}\right)^2\right]^{1/2} \qquad\qquad (7.122)$$

and

$$\delta = \left[\left(1 - \frac{\lambda_{0r}}{b}\right)^2 + \left(\frac{\lambda_{0i}}{\rho b}\right)^2\right]^{-1/2} - 1, \qquad (7.123)$$

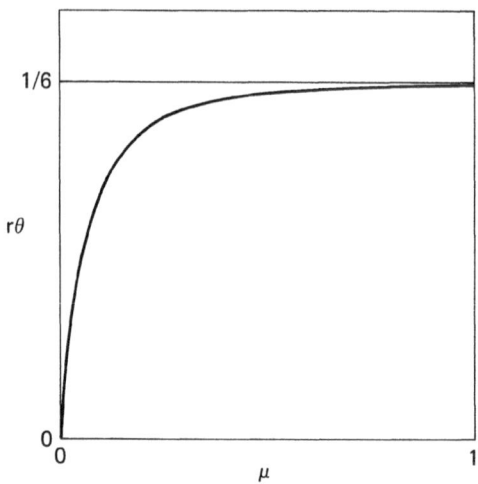

Fig. 7-8 Optimal tuning parameter as determined from Eq. (7.128).

so that

$$\delta \approx \frac{\lambda_{0r}}{b} - \frac{1}{2}\left(\frac{\lambda_{0i}}{\rho b}\right)^2. \tag{7.124}$$

The assumption $\delta \ll 1$ requires both terms on the right-hand side of (7.124) to be small. Maximizing (7.121) with respect to ρ then gives

$$\theta^3 + \mu^2(3\theta - \tfrac{1}{2}) = 0, \tag{7.125}$$

where

$$\theta = \frac{\lambda_{0i}^2}{4\rho^2 b\lambda_{0r}} \tag{7.126}$$

and

$$\mu = \frac{1}{4}\left|\frac{\lambda_{0i}}{\lambda_{0r}}\right|. \tag{7.127}$$

Equation (7.125) has only one real root since its θ–derivative is never zero for real θ. This root is

$$\theta = (\tfrac{1}{4}\mu)^{1/3}\{(\sqrt{16\mu^2 + 1} + 1)^{1/3} - (\sqrt{16\mu^2 + 1} - 1)^{1/3}\}. \tag{7.128}$$

As μ increases from zero, θ increases monotonically from zero to its limiting value of $\tfrac{1}{6}$. This is shown in Figure 7-8.

Maximizing (7.121) with respect to b is equivalent to maximizing (7.123). The result, invoking (7.126), is

$$b_{\max} = \frac{\lambda_{0r}}{1 - 4\theta} \lesssim 3\lambda_{0r}. \tag{7.129}$$

However, this value is unacceptably small since many eigenvalues of high mode numbers would fall outside the ellipse. Thus, we take the smallest allowable value for b:

$$b = -\frac{4}{\delta x^2}. \tag{7.130}$$

This excludes from the ellipse the "odd-even" mode $(n_x, n_y) = (\frac{1}{2}N_x, \frac{1}{2}N_y)$. This choice allows an effective doubling of the computational efficiency (to be demonstrated below) at the expense of having to perform a single follow-up iteration to eliminate the "odd-even" mode:

$$\psi \rightarrow \psi + \frac{1}{8}\delta x^2(\mathcal{L}\psi - S). \tag{7.131}$$

From (7.114) and (7.130) the midpoint of the ellipse, b, is just equal to the center coefficient of the finite-difference operator $\mathcal{L}$ in (7.114). The recursion formulas for the iterative solution, (7.90) or (7.97) as appropriate, involve the operator $\mathcal{L} - b$, whose central coefficient is zero. This results in a natural separation of the finite-difference solution into odd and even components, separated according to whether the sum of the indices $i + j$ is odd or even. In fact this odd even separation extends to an arbitrary number of spatial dimensions. This separation would allow one to define ψ^n for even (odd) n only on even (odd) gridpoints. One could then refine the solution in a "hopscotch" fashion, updating only half the points at a given iteration.

We have not taken advantage of this odd-even grid separation in any of the results presented here. An interested user should keep in mind two caveats: if the eigenvalue ellipse is prolate $(c/a > 1)$, convergence is not guaranteed for odd-numbered iterates [see Eq. (7.88)]; and to overcome the inherent fetch-speed limitation of a pipeline computer, one must separate even and odd grid point quantities into contiguously stored arrays.

It remains to give expressions for λ_{0r} and λ_{0i} for the model problem. Defining

$$\sigma = (s_x^2 + s_y^2)_{\min}^{1/2} \approx \frac{2\pi}{N_x} \tag{7.132}$$

(for periodic boundaries), (7.119) gives for $\sigma \ll 1$

$$\lambda_{0r} \approx -\frac{\sigma^2}{\delta x^2},$$

$$\lambda_{0i} = \frac{\bar{a}\sigma}{\delta x^2}. \tag{7.133}$$

In (7.132), "min" refers to a minimum overall wavenumbers excluding the $(0,0)$ mode. Ellipse parameters a and b are now specified by (7.122) and (7.130). We obtain c from (7.126) and (7.103):

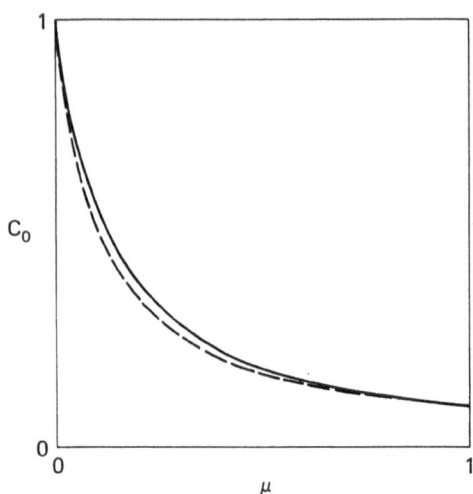

Fig. 7-9 Normalized convergence rate (7.137) (solid); approximation (7.138) (dashed).

$$c = \frac{a\lambda_{0i}}{\sqrt{4\theta b\lambda_{0r}}}. \tag{7.134}$$

Taking into consideration (7.106), (7.109), (7.110), and (7.129), we have the approximate convergence rate

$$C \approx \sqrt{2\delta + \rho^2} - \rho. \tag{7.135}$$

We can use (7.127), (7.128), and (7.133) to express (7.135) in terms of μ. The result is

$$C \approx 2^{-1/2}\sigma C_0(\mu), \tag{7.136}$$

where

$$C_0(\mu) = \sqrt{2}[\sqrt{\tfrac{1}{2} - \theta + \mu^2/\theta} - \mu\sqrt{\theta}]. \tag{7.137}$$

Recall that $\sigma \approx 2\pi/N_x$ for doubly periodic boundaries. The dependence of C_0 upon μ is shown in Fig. 7-9. Also plotted in Fig. 7-9 (dashed line) is the approximation

$$C_0(\mu) \approx (9.5\mu + 1)^{-1}. \tag{7.138}$$

Using (7.128), we find the following limiting forms for C:

$$\begin{aligned}
C &\approx \sigma 2^{-1/2}(1 - 3(\tfrac{1}{2}\mu^2)^{1/3}), & \mu \ll 1, \\
C &\approx \sigma\mu^{-1}6^{-3/2}, & \mu \gtrsim 1.
\end{aligned} \tag{7.139}$$

This shows that as μ increases from 0 to 1, the convergence rate drops by approximately a factor of 10.

We can see from (7.132) and (7.136) that the number of iterations required to reach a certain level of error reduction is proportional to N_x. Thus, the total number of operations required is proportional to $N_x^2 N_y$. Therefore, it is advantageous to correct the long-wavelength components of the solution on a coarse mesh and interpolate onto a fine mesh before completing the solution. The importance of regridding as a means of accelerating convergence of explicit relaxation schemes has been recognized for some time. The interpolation between coarse and fine meshes introduces truncation error into the solution, so that the error estimates derived earlier may not hold on the fine mesh. For this reason, it is best to alternate between fine and coarse meshes more than once, attempting only a modest error reduction with each pass. The procedure to be described below was developed through practical experience in simulating physical systems. The details of this procedure can be justified by a simple model given by McDonald (1980). In practice, use of a coarse grid becomes important for $N_x \gtrsim 50$. For a grid of 128×128 interior points, a coarse grid of 32×32 interior points is used. First, the residual $\mathscr{L}\psi^0 - S$ is extracted on the fine grid. Then this residual and **A** are defined on the coarse grid using block averages of 16 fine grid points per coarse grid point. A coarse grid potential correction is initialized to zero and a large number of iterations are performed. Typically, we use three times as many iterations on the coarse grid as on the fine grid. Then the potential correction is defined on the fine grid using bilinear interpolation, and the result is subtracted from ψ^0. Then iterations are performed on the fine grid, making no attempt to make significant improvements in the lowest 5–10 modes of the solution. That is, we arbitrarily increase σ in (7.132) by a factor of 5–10. This results in improved convergence of the higher modes. The effectiveness of this fine grid–coarse grid approach may be improved significantly in a time-dependent problem by two-level extrapolation for the trial solution ψ^0.

Numerical results: As a test of the iterative procedure (7.90) and the convergence estimate (7.136) we have solved (7.111) on meshes of 32×32 and 48×48 interior grid points without regridding; and on a 128×128 mesh with a reduced mesh of 32×32 interior points. All tests used doubly periodic boundary conditions. The numerical convergence rate comparisons were carried out as follows. Arbitrary forms were adopted for **A** and a reference solution Ψ. Then a source term was generated from Ψ numerically using the difference operator (7.114). This source term was used in the iteration (7.97), with the approximate solution being initialized to zero. After a large number N of iterations (usually $N = 40$), a relative error E was defined to be the root mean square residual of (7.111) divided by the root mean square of the source, S. The average convergence rate was then taken to be $-(\log E)/N$.

The routines used in the Chebychev solver are listed as Appendix E. In all cases convergence was fast enough to be consistent with (7.136), and

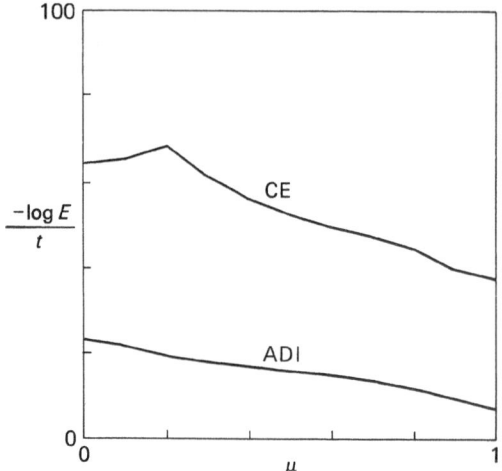

Fig. 7-10 Computational efficiencies of the Chebychev explicit method (CE) and Alternating Direction–Implicit method (ADI) when applied to a series of numerical tests in which $|A|$ and thus μ vary from case to case. Plotted is the logarithm of the error reduction factor after many iterations, divided by the elapsed computer time in seconds.

in most cases was faster than (7.136). The one case in which convergence was just equal to (7.136) for all μ was for $\Psi_{ij} = \sin 2\pi i/N_x$, and $A_x = \text{const}$, $A_y = 0$.

Figure 7-10 shows convergence rates per second of computer time, $-\log E/t$, obtained using the two-pipe NRL Texas Instruments ASC to solve the same set of problems with the Chebychev Explicit method (CE) (upper curve) and with an Alternating Direction Implicit method (ADI) (lower curve). These comparisons were made on a 50×50 mesh using data from the EJET plasma turbulence code (McDonald *et al.*, 1975). The vector A was computed from a turbulent plasma density distribution $n_e(x, y)$ having a spectral power index of approximately -3.5:

$$A_x = f \frac{\partial n_e}{\partial y},$$

$$A_y = -f \frac{\partial n_e}{\partial x}. \tag{7.140}$$

The proportionality constant f was adjusted to give a desired value of μ [see (7.127)] from 0 to 1. Figure 7-10 summarizes 11 separate tests with A generated from the same n_e, but scaled to give 11 equally spaced μ values. For these tests Ψ was the electrostatic potential appropriate to n_e.

The ADI solution used logarithmically spaced iteration parameters and a partially vectorized tridiagonal solver. Although ADI converged faster than CE per iteration, the limited vectorizability of the method resulted in slow execution. Thus, the result in Fig. 7-10 is machine dependent. The execution times per iteration were 25.8 ms for ADI and 1.32 ms for the CE method. To illustrate the computational efficiency of the explicit method, the lower limit on execution time per iteration (neglecting overhead and boundary value resetting) would be (48 × 48 interior points) × (11 operations per point) × (40 ns per operation) = 1.01 ms. Thus, the explicit method runs at 76% efficiency on a modest 50 × 50 mesh. ADI could be made more competitive by the use of cyclic reduction rather than tridiagonal solution in the integration direction, but it is unlikely that any improvement would bring the convergence rate per second up to that of the explicit method. In addition, the boundary conditions are much easier to change in the explicit method than in the implicit one. Recall, too, that when the eigenvalue ellipse is oblate, one can use "hopscotch" updating and achieve, in principle, a doubling of the explicit method's efficiency.

Another comparison was made using a recently proposed method employing the Conjugate Gradient (CG) algorithm. This particular CG method requires inversion of the self-adjoint part of $\mathscr{L}$ [i.e., $\frac{1}{2}(\mathscr{L} + \mathscr{L}^T)$] once per iteration. For this problem an optimized complex fast Fourier transform (FFT) is used to invert the Laplacian operator. The results of the comparison are not shown in Fig. 7-10, but are comparable to the ADI results for $\mu \geq 0.1$. For $\mu = 0$ the method is direct and thus superior to the others. However, for $\mu \geq 10^{-4}$ the convergence of CG is no greater than that of CE. Use of a fully optimized real transform would increase the overall speed of CG by approximately a factor of 2, leading to results intermediate to the two curves of Fig. 7-10. However, this particular CG method is not as generally applicable as ADI or CE since it may become unwieldy when the self-adjoint portion of $\mathscr{L}$ contains varying coefficients.

CHAPTER 8

Vectorization of Fluid Codes

Each of the computational techniques described in this volume is designed to run efficiently on a vector computer. This chapter describes the principles underlying their implementation on the Texas Instruments Advanced Scientific Computer (ASC).

On fast computers, superficially minor details of the programming can change the computer time needed to solve a given problem by a large factor. Table 8-1 demonstrates the possibilities on the ASC. A two-pipe ASC is capable of performing 1000 additions in 646 cycles (52 μs). If the operation is coded differently, however, it requires 22437 cycles (1800 μs). The second version, which runs 35 times slower, performs no more mathematical operations; it simply does not take advantage of "parallelism."

"Parallelism" is present in computers as old as the IBM 360/91 and CDC 7600. Advertisements of a vector processor, loop mode, cache memory, or fast memory buffer identify computers which have such parallelism. The TI ASC, Cray Research CRAY-1, CDC STAR, and CYBER 203 all make extensive use of it. All the hydrodynamics algorithms described in this volume can benefit from program design which makes use of parallelism in these machines.

What is distinctive about a computer with parallelism? It has an arithmetic unit which is faster than its memory access. In the fastest of these computers, the limit on memory access is the speed of propagation of an electrical impulse in a wire; that is, propagation at a fraction of the speed of light. In others,

Table 8-1 Comparison of individual (scalar) and group (vector) processing times in cycles for the operation.

Length of DO LOOP	Scalar Time	Vector Time	
		1 Pipe	2 Pipes
1	60	91	92
10	257	104	113
50	1171	149	139
100	2288	203	170
500	11244	615	402
1000	22437	1154	646

the limit is that the memory itself is slow. In all of them, access to memory is speeded up by sending several words to or from memory as a group, in a time substantially less than would be required to send them individually.

The group transfer of words to and from memory has one very large advantage—speed. This is clearly indicated in Table 8-1. It also has two serious difficulties—complexity and causality. The complexity problem should be obvious. The arithmetic unit must process whole groups of data, rather than individual words. The causal problem is more subtle. The arithmetic processing of one group may require access to another group which is in transit to or from memory, and therefore unavailable. Waiting for individual words to make round trips to memory thus interrupts the group processing of data.

Section 8.1 describes the group transfer process on several fast computers to show how they are similar. Section 8.2 explains how this process is used, with Fortran examples. Section 8.3 demonstrates the causality problems, and explains how to avoid many potential problems by careful choice of algorithms. Section 8.4 illustrates these general principles with specific examples drawn from the programming of the hydrodynamics algorithms presented in the preceding chapters. Finally, Section 8.5 summarizes the general principles of exploiting parallelism.

8.1 Speed in Hardware

The basic limit on computing speed is the propagation speed of an electrical impulse in a circuit, approximately 20% of the speed of light. Dense integrated circuit elements in the arithmetic unit, and in the complete central processing unit (CPU), can now perform arithmetic operations in tens of nanoseconds or less. The physical volume of a large memory is such that the distance from its center (the logical location for its CPU) to its periphery is seldom less than 3 m, and usually much greater. But 3 m of propagation implies a 50-ns signal transit time, or a 100-ns round trip (request for a memory word, followed by return of that word). Thus, the speed limit in such a computer is enforced by memory access delays.

The detour around this speed limit is obvious: Fetch more than one word with each memory request. For example, suppose a hypothetical CPU (call it HYPPO) can complete one operation in 10 ns, and a memory access requires 100 ns to fetch any number of words. Then HYPPO should fetch at least 10 words with each memory reference. On closer examination, the number is actually larger. Suppose HYPPO can add 10 pairs of numbers, $c_i = a_i + b_i$, in 100 ns. Then a large number of pairs can be added without loss of CPU efficiency only if, during each 100-ns period, the CPU contains 60 words in buffers. Half of them are being used by the CPU, 10 c_i values are being sent to memory from the last 10 operations, and 20 a_i and b_i values are being obtained from memory for the next 10 operations. All these buffers, plus the associated memory drivers and logic to switch the CPU among these buffers, provide the other reason this hypothetical computer is named HYPPO. It is big.

Most fast computers are designed so that group or pipeline access to memory occurs when arithmetic operands are stored sequentially in memory. That is, if the three arrays A_i, B_i, and C_i are arranged so that their elements are stored in consecutive memory locations, then the IBM 360/91, CDC 7600 and STAR, and TI ASC can fully utilize their CPU speeds. The CRAY-1 has more general memory access; it can quickly move groups of data to or from memory whenever their memory addresses have a constant displacement; e.g., every second word, or every tenth word.

In each fast computer, there is an "overhead" or startup time associated with group processing. As a result, groups must have a minimum length, typically 5–20, in order to make group processing more efficient than processing individual elements. This minimum size varies significantly among computers.

On fast computers, many choices must be made in the development of an efficient program. The data layout must be carefully organized. How (or whether) to use group operations is often decided by global properties of the algorithm being implemented. Each programmer is faced with the necessity of spending several years learning the intricacies of very specialized hardware. Fortunately, computer manufacturers are now offering a viable alternative in the form of high-level language compilers (Fortran, Pascal, etc.), which will automatically or semiautomatically provide access to this hardware speed. Use of such a compiler is the subject of the next section.

8.2 Speed in Fortran

The state of the compiler art is high. Many Fortran compilers are adept at recognizing group actions, such as DO loops, and taking special action to execute them efficiently. For the ASC, this means generating "vector" instructions. Let us see how this generation is related to the Fortran source text.

We shall begin with the addition of 100 pairs of numbers, $C_i = A_i + B_i$. The Fortran text reads:

        DO   1   I = 1,100

        C(I) = A(I) + B(I)

    1   CONTINUE

A traditional "scalar" computer executes five assembly language instructions 100 times. There are two memory fetches (A_i and B_i), one addition, one store to memory (for C_i), and an instruction that increments a counter, tests, and branches back to load the next pair of input operands. Thus 500 scalar instructions are executed to add arrays A and B.

The ASC can generate such scalar code (we shall see why in the next section) or it can generate object code which executes very differently. The vector code for adding the 100 pairs of operands consists of a single assembly language instruction and an associated table, built by the Fortran compiler. The table contains the starting address in central memory of the input and output arrays and the increments for stepping through the arrays (in this case, unity). A vector instruction executes by continuously streaming operands from memory into the central processor (CP), where the addition takes place, and continuously streaming answers back to memory. The vectorized addition, in this example, may be thought of as 100 additions simultaneously occurring on the 100 pairs of input operands. Actually, during execution, some elements of A and B are being read from central memory, some elements of A and B are undergoing addition in the CP, some answers (array C) are in output buffers, and some are being written to memory.

Table 8-1 lists the time that it takes to execute a DO loop which performs various numbers of additions on the ASC in scalar mode and in vector mode. This particular ASC has two arithmetic units (AU's), also called pipelines or pipes. The additions can be performed by one of them, or the A, B, C arrays can be split in half, and each AU can do half the additions. Vectorized times are given for utilization of one or both AU's. Times are given in CP clock cycles; a cycle is 80 ns.

Notice that, for scalar execution, the time per addition remains around 24 cycles whether 10 or 1000 additions are performed. In two-pipe vector mode, the time per addition decreases from 11 cycles to 0.65 cycles as array size increases from 10 to 1000 elements. This timing pattern is characteristic of all vector operations on the ASC. When arrays are large, the benefit from vectorization and from using two arithmetic units becomes very substantial. Thus, for most hydrodynamics codes, increasing array sizes or refining grid resolution is much less costly in computing time if the code is run on a vector computer than if the code is run on a scalar computer.

For calculations involving multiply dimensioned arrays, the ASC differs from other vector computers. Loops nested two or three levels deep, containing operations on doubly or triply subscripted arrays, also may be collapsed into a single vector instruction. The loop

```
     DO  2  K = 1,10
     DO  2  J = 1,10
     DO  2  I = 1,50
     C(I,J,K) = A(I,J,K) + B(I,J,K)
  2  CONTINUE
```

Table 8-2 Partial list of the more than 100 vector operations in the ASC hardware. Starred entries have vector functions for half-word, integer, real and double precision arguments.

*Add (Algebraic)	*Min
*Add (Magnitude)	*Multiply
*Compare ($<$, $\leq$, $>$, $\geq$, $\neq$)	Normalize
*Divide	*OR
*Dot Product	Order
*Exclusive OR	Peak Pick (Max and Min)
Fix-to-Float	*Replace
Float-to-Fix	Search
*Map	*Select
*Max	*Subtract (Algebraic)
Merge	*Subtract (Magnitude)

executes as a single vector instruction which adds 5000 pairs of numbers. No other computer can make such a triple loop into a single instruction, unless the upper limits of the loops are equal to the corresponding dimensions of all the arrays, so that the triple loop is really equivalent to a single loop. Most other fast computers will optimize or vectorize only the inner loop.

Useful computations are not restricted to simple additions. The typical DO loop in a program contains several operations, both arithmetic and logical. Here again the compiler should recognize (and the ASC compiler usually does) that the more complex operation is a series of single array operations. Thus the DO loop

DO   3   J = 1,M

DO   3   I = 1,N

$Y(I,J) = A(I,J) * X(I,J) ** 2 + B(I,J) * X(I,J) + C(I,J)$

$P(I+J) = Q(I | 5 * J - 4) + R(I,J)$

3   $S(I,J) = AMIN1(T(I,J), U(I,J))$

can become 7 vector instructions (or 14, if they are split over two AU's).

An arithmetic expression must meet two criteria to be vectorizable. First, its dependence on the DO-loop indices must be linear. Second, the hardware must have the desired operation implemented (see Table 8-2). There are two circumstances, however, where the required decisions are more complex. They arise where the arrays are not independent and where conditional execution (an IF statement) is involved. These are the subjects of the next two sections.

8.3 Problems with Causality

Physicists invoke causality to discuss events which occur simultaneously, or nearly simultaneously, at different locations. Causality problems occur in a computer when the operands of a series of operations are interdependent. If the programmer attempts to use parallelism, he finds that some results are needed in one place while they are simultaneously being calculated in another. One of the two following DO loops has problems with causality.

$$\text{DO} \quad 4 \quad I = 1, 99$$
$$4 \quad A(I) = A(I) + A(I+1)$$
$$\text{DO} \quad 5 \quad I = 2, 100$$
$$5 \quad A(I) = A(I) + A(I-1)$$

Which one has the problem?

In vector processing of an addition, either physically distinct hardware is performing successive additions [that is, one AU is doing the ith addition while another is doing the $(i + 1)$th], or the complete addition (address decoding, operand fetch, addition, store, etc.) is performed in steps, and the ith addition is in one step while the $(i + 1)$th is in the step behind it. In either case, the result of the ith addition is not available until after the $(i + 1)$th is already underway. Thus, in the example given above, statement number 5 has causality problems.

The compiler will vectorize statement 4, but not statement 5. If a statement like 5 occurs only infrequently in a program, it is reasonable to let it be performed in scalar operations. However, if its execution dominates the computation time, it must be rewritten so that it can be vectorized. The easiest way to do this begins with an examination of the desired results. They are

$$B(2) = A(1) + A(2),$$
$$B(3) = A(1) + A(2) + A(3),$$
$$B(4) = A(1) + A(2) + A(3) + A(4),$$

etc. Here B is used to represent the A array after execution.

The most obvious method is to directly vectorize these cumulative sums. The operation in statement 5 may be vectorized as

$$\text{DO} \quad 6 \quad I = 2, N$$
$$B(I) = 0$$
$$\text{DO} \quad 6 \quad J = 1, I$$
$$6 \quad B(I) = B(I) + A(J)$$
$$\text{DO} \quad 7 \quad I = 2, N$$
$$7 \quad A(I) = B(I)$$

Statement 6 vectorizes on J, and statement 7 also vectorizes. Note, however, that the operation count has been increased from N to $\frac{1}{2}N(N + 1)$. Statement 5 has been vectorized, but it will take longer to execute than statement 4.

The most efficient method of vectorizing statement 5 lies between the original form and this one. Suppose we rewrite the operations on the B's in terms of alternate A's. That is,

$$B(2) = A(1) + A(2),$$
$$B(4) = A(3) + A(4) + B(2),$$
$$B(6) = A(5) + A(6) + B(4).$$

Now we must perform $2N$ additions, but we have a scalar operation on only $\frac{1}{2}N$ B's. This scalar opeation can, in turn, be replaced by a similar process on the B's, at the cost of $2(\frac{1}{2}N)$ vector additions plus $\frac{1}{4}N$ scalar operations. This is a recursive prescription for reducing the length of the scalar operation until its computational cost is insignificant.

This process is a form of cyclic reduction, closely related to that employed in the direct Poisson solvers discussed in Section 7.1. For large N, the N scalar operations in statement 5 are replaced by $N \ln N$ vector operations, but with a great increase in computational complexity.

There are many other cases in which apparently purely scalar operations may be vectorized. Fortunately, in most of the important cases, a subroutine has been written to perform that operation. Instead of coding, one carries out a library search for the appropriate subroutine.

For practical purposes, these causality problems occur in one dimension. If the data dependence is in only one of several dimensions, for example, if the source code is

```
DO  8   I = 1, N
DO  8   J = 2, N
8   A(I,J) = A(I,J-1) + B(I,J)
```

then an interchange of the two "DO 8" statements is enough to allow the inner loop (now on I) to vectorize. Perhaps more importantly in the example

```
DO  9   I = 2, N
DO  9   J = 2, N
DO  9   K = 2, N
9   A(K,J,I) = A(K-1,J-1,I-1) + B(I,J,K)
```

cyclic reduction is needed only for the innermost loop. When the loop on K has been vectorized, accesses to the "dependent" data in the $J - 1$ and $I - 1$

planes are far removed in time from the J and I accesses (during succeeding cyclic reductions), so no causality problem occurs.

This section and the last one have dealt mostly with simple DO loops. Let us now look at important parts of the algorithms we have seen in other chapters.

8.4 Examples

The vectorization of a single arithmetic operation is conceptually straightforward. The execution speed of such operations is also very impressive, when compared with conventional scalar speeds. However, the successful vectorization of a single arithmetic operation is not enough to solve a real problem.

Does vectorization apply to practical problems? The most productive way to answer such a question is to pick a problem and try vectorizing it. Of course, the specific problem chosen will greatly influence the probability of success. One might choose the problem of multiplying two matrices and get a solid "yes." Alternatively, one might choose "sorting a list of names" and get an equally solid "no."

We are concerned in this volume with fluid dynamics problems, so we shall define a "practical problem" to be one which performs a substantial part of a fluid calculation. In this section we will discuss two examples—one which is mostly arithmetical and one which involves logical choices as well.

The first example is the elliptical partial differential equation solver based on explicit Chebychev iteration which was described in Section 7.3. Appendix E lists the core subroutine EXCIT and two subroutines it calls in obtaining the solution. Most of the code consists of arithmetic statements, written so as to make efficient use of the ASC architecture. Here we will highlight the features incorporated in the code for this purpose.

The first noteworthy features are seen in some of the nonexecutable Fortran statements. Each subroutine contains PARAMETER, REAL, and COMMON statements which give the Fortran compiler information about the way in which data will be used. The PARAMETER statements indicate that certain Fortran variables (such as MX and MY) are actually constants. Only constants or parameters such as these may appear in the dimensions of a COMMON statement.

The way the REAL statements in the subroutines use parameters reflects a problem of vector computers. For example, the subroutine argument P is dimensioned as P(NX/MX/, NY/MY/), which tells the Fortran compiler two things. First, on any execution of the subroutine, the dimensions will be P(NX,NY), with whatever values NX and NY have at subroutine entry. Second, the values of NX and NY will always satisfy NX $\leq$ MX, NY $\leq$ MY.

This "maximum dimension" information is used to allocate temporary storage, when necessary. The scalar statement

$$A = 3*B - 2*C$$

is executed by performing one multiplication, storing the result (or saving it in a register), performing the second multiplication, then subtracting the two results, and storing the difference. A similar vector statement

DO   10   J = 1, NY

DO   10   I = 1, NX

10   S(I,J) = P(I,J) − 2. * Q(I,J)

requires the use of temporary storage for the result of the multiplication, before the subtraction can be performed. The amount of storage needed is NX * NY words, but this can dynamically vary at different calls to the subroutine, with different values of NX and NY. The ASC Fortran compiler uses the information in the REAL statement to determine the amount of storage (in a private COMMON block) to reserve for "temporary" vector storage. In this case, it would reserve MX * MY words. For a more complicated expression it would reserve this amount multiplied by the maximum number of temporary results needed during evaluation of the expression.

Virtually all of the code in SUBROUTINE EXCIT has been vectorized, but the part which has received the most careful attention is in the middle of the subroutine. The statements with labels 32 and 36 consume most of the CPU time, in typical applications. Several things have been done to make execution efficient. The array subscripts are ordered so the first subscript is always varied by the innermost DO statement. The subscripts are all linear functions of the loop indices. The variable before the equal sign does not appear after it, thus avoiding even the appearance of a vector hazard. Of course, the call to subroutine PBCSET is outside the loop, and is made only once to perform operations on a complete array of data.

This example showed that a nontrivial physics problem can be solved without logical statements in the innermost loops. It is instructive to see how a problem with logical choices in the underlying algorithm is vectorized. The ETBFCT algorithm (Section 3.2) is a particularly good illustration. It solves the central problem of computational hydrodynamics, that of the convective equation. Equally important, it has array, scalar, and conditional operations. Thus, it is not obvious whether it should be efficient on a vector computer.

The ETBFCT algorithm itself was described in Section 3.2. The first three of its stages (transport, finding diffusive fluxes, and finding antidiffusive fluxes) and the last (calculation of the flux-limited solution) are linear operations on arrays of fluid quantities. Thus, they are easy to vectorize.

The remaining operation in ETBFCT [the flux-limiting procedure (3.25)] is defined by partial operations in various special cases, determined by inequalities. We have seen that expressions involving IF tests do not vectorize. Thus, vectorizing the flux limiter requires casting it in a form which does not explicitly test for these inequalities. In order to see how this is done, let us look at the actual limiting operations.

The purpose of the limiter is to ensure that no new maxima or minima are formed. In principle, this means comparing each set of three adjacent values of variables, at successive time steps, determining whether a new maximum or minimum has formed, and taking corrective action if necessary. The substantial number of comparisons involved can be reduced by focusing on the element of transport—the flux.

The flux $F_{i+1/2}$ transports density from the cell with density ρ_i to the cell with density ρ_{i+1}. This transport creates a new maximum or minimum only if it reverses the gradient before it $(\rho_i - \rho_{i-1})$ or after it $(\rho_{i+2} - \rho_{i+1})$. Examination of the flux and density gradients therefore apears to reduce the required number of tests to two for each cell. The next step is to eliminate the tests. This can be done if the unlimited flux and the result of flux limiting can together be expressed in terms of Fortran library functions.

The action of flux limitation consists of reducing the calculated flux to a value which is small enough that it does not reverse adjacent gradients. If $F_{i+1/2}$ is positive, that means (in Cartesian coordinates)

$$F_{i+1/2}^{c+} = \min(F_{i+1/2}, \tilde{\rho}_{i+2} - \tilde{\rho}_{i+1}, \tilde{\rho}_i - \tilde{\rho}_{i-1}),$$

since $F_{i+1/2}$ will subtract from ρ_i and add to ρ_{i+1}. If $F_{i+1/2}$ is negative, the gradients must be limited in the opposite direction:

$$F_{i+1/2}^{c-} = \min(F_{i+1/2}, \rho_{i+1} - \rho_{i+2}, \rho_{i-1} - \rho_i).$$

These two expressions can be combined with the absolute value and the signum function

$$\operatorname{sgn}(x) = \begin{cases} 1, & x \geq 0 \\ -1, & x < 0 \end{cases} \tag{8.3}$$

in the form

$$F_{i+1/2}^{c\pm} = \sigma_{i+1/2} \min[\operatorname{abs}(F_{i+1/2}), \sigma_{i+1/2}(\tilde{\rho}_{i+2} - \tilde{\rho}_{i+1}), \sigma_{i+1/2}(\tilde{\rho}_i - \tilde{\rho}_{i-1})], \tag{8.4}$$

where

$$\sigma_{i+1/2} = \operatorname{sgn}(F_{i+1/2}). \tag{8.5}$$

Even this is not quite enough, since the limiting process should not be allowed to reverse the flux. Thus, the final limited flux is determined by

$$F_{i+1/2}^{c} = \sigma_{i+1/2} \max\{0, \min[\operatorname{abs}(F_{i+1/2}), \sigma_{i+1/2}(\tilde{\rho}_{i+2} - \tilde{\rho}_{i+1}),$$
$$\sigma_{i+1/2}(\tilde{\rho}_i - \tilde{\rho}_{i-1})]\} \tag{8.6}$$

is in planar geometry, where the curvilinear scale factor equivalent to (3.25) $\Lambda_{i+1/2} = 1$. This scalar formula for the limited flux $F_{i+1/2}$ can be expressed entirely in terms of standard Fortran library functions which have been vectorized.

Because the SIGN operation is not expanded in line on the ASC, the limiter (8.6) is evaluated as shown in Appendix A. We define the real quantities MASK1, MASK2, and MASK3 via a data statement. MASK1 consists of a full word with 1 in the first (sign) bit and zeros thereafter, MASK2 is set equal to (floating point) unity, and MASK3 has 1 in each bit location. Loop 21 ANDs the difference $\tilde{\rho}_{i+1} - \tilde{\rho}_i$ with MASK1, yielding the sign of the latter in the first bit and zeros in the others. Loop 22 ORs this quantity with MASK2 to obtain $\sigma_{i+1/2}$, then ANDs MASK3 with the raw flux $F_{i+1/2}$, yielding the absolute value of the latter. Loops 23 and 25 multiply $\tilde{\rho}_{i+2} - \tilde{\rho}_{i+1}$ and $\tilde{\rho}_i - \tilde{\rho}_{i-1}$, respectively, by $\sigma_{i+1/2}$ and the appropriate curvilinear scale factors. Loops 24 and 35 calculate the minimum specified in (8.6), then 36 calculates the maximum and 37 reapplies the sign $\sigma_{i+1/2}$. A single executable statement (8.6), which assigns the value of a complex arithmetic expression to the array containing the fluxes $F^c_{i+1/2}$, has been reexpressed using no fewer than 10 loops. The resulting code executes up to two orders of magnitude faster than the original scalar version!

This completes the vectorization of the ETBFCT algorithm, but one more subject deserves explicit mention. Boundary conditions have not been discussed. When calculations on the interior of the grid are fully vectorized, a complicated boundary condition sometimes becomes the most time-consuming part of a calculation.

The ETBFCT module, as implemented, incorporates the boundary condition in the vectorized subroutine, provided it is one of three general types. It treats the simplest, constant boundary values by including the boundary values in the calculation, but only updating the interior values. Reflecting (or impermeable) boundary conditions are treated by placing a "guard" cell beyond the outermost cell and calculating the values there in such a way that the midpoint between the last cell and the guard cell is a reflection point. Finally, periodic boundary conditions are incorporated, again with a guard cell, by assigning to the guard cell the value at the other end of the grid. These three common boundary conditions are thus incorporated into the algorithm without degrading the vector performance of the algorithm.

8.5 Summary of Parallelism Principles

There are a number of optimization principles that apply to most fast computers. These have applicability to general vectorization efforts both in and out of fluid dynamics. The methods are straightforward, but must be applied many times, and often with considerable ingenuity, to completely vectorize a code. Programs which are coded following these principles will generally run faster than programs which are not, provided that an optimizing compiler is used. The factor may be two or more on machines such as the IBM 360/91 or the CDC 7600, or it may be as much as a factor of 10 or 100 for a vectorizable code on a vector computer. The corresponding cost is usually an increase in the computer memory required for data (and arrays of intermediate results in vector computations).

First, certain principles apply to any fast computer. They are:

1. Plan programs and subroutines to operate on arrays of data rather than on individual elements.
2. Store the arrays in memory so that the elements are usually read from or stored into consecutive memory locations.
3. The DO loop index (the innermost one, if there is more than one) should be the *first* subscript of an array wherever possible.
4. If a scalar appears on the left-hand side of an assignment in a DO loop, replace it by an array.
5. Remove IF statements from DO loops wherever possible.
6. Use optimized or vectorized subroutines to eliminate IF statements or replace code which the compiler does not optimize.

Note that principle 2 usually means maximizing the application of principle 3. Principle 6 applies to IF statements which are performing a function such as MIN, MAX, or a cyclic reduction or merge.

Second, there are principles which relate to computers which have more than one arithmetic unit. The application of these principles is fairly machine dependent. For pipeline machines with more than one pipeline (TI ASC, CYBER 203) operations on independent data streams should be made adjacent:

7. Write short Fortran lines.
8. Code independent pairs of array operations in neighboring lines.

Principle 7 also applies to loop-mode machines like the IBM 360/91, where DO loops must be short enough for the required instructions to fit into the "loop stack." For machines which have independent functional units (adder, multiplier, etc.) such as the CDC 7600 and CRAY-1, the principles are

9. Write long Fortran lines.
10. Code expressions so that independent operations alternate on dependent data.

Beyond these 10 principles, it is necessary to consider the detailed architecture of the target machine. However, codes which follow these principles should require nearly the minimum amount of restructuring in order to take advantage of the detailed hardware features.

The codes listed in the Appendices are efficient not only because of the solution algorithms they employ, but also as a result of the degree to which they have been vectorized. Careful study of these examples will reveal a great deal regarding both the general principles of vectorization and those techniques peculiar to the ASC.

Appendix A

```
 1*         SUBROUTINE ETBFCT (RHOO, RHON, N, RBC, LBC)
 2* C
 3*            REAL     RHOO(N), RHON(N)
 4* CD
 5* CD    * * * * * * * * * * * * * * * * * * * * * * * * * * * * * * * * *
 6* CD
 7* CD    ETBFCT (RHOO, RHON, N, RBC, LBC)                         D.3
 8* CD    ORIGINATOR - J.P.BORIS         CODE 7706, NRL        OCT.1975
 9* CD
10* CD    DESCRIPTION:  THIS ROUTINE SOLVES GENERALIZED CONTINUITY EQUATIONS
11* CD    OF THE FORM        DRHO/DT = -DIV (RHO*V) + SOURCES      IN EITHER
12* CD    CARTESIAN, CYLINDRICAL, OR SPHERICAL GEOMETRY. THE FINITE-DIFFER-
13* CD    ENCE GRID CAN BE EULERIAN, SLIDING REZONE, OR LAGRANGIAN AND CAN
14* CD    BE ARBITRARILY SPACED. THE ALGORITHM USED IS A LOW-PHASE-ERROR FCT
15* CD    ALGORITHM, VECTORIZED AND OPTIMIZED FOR SPEED. A COMPLETE DESCRIP-
16* CD    TION APPEARS IN NRL MEMO REPORT #3237, "FLUX-CORRECTED TRANSPORT
17* CD    MODULES FOR SOLVING GENERALIZED CONTINUITY EQUATIONS".
18* CD
19* CD    ARGUMENTS:  IN THIS ROUTINE THE RIGHT BOUNDARY AT RADR IS HALF A
20* CD    CELL BEYOND THE LAST GRID POINT N AT RADN(N) AND THE LEFT BOUNDARY
21* CD    AT RADL IS HALF A CELL BEFORE THE FIRST GRID POINT AT RADN(1).
22* CD    RHOO    REAL ARRAY(N)     GRID POINT DENSITIES AT START OF STEP   I
23* CD    RHON    REAL ARRAY(N)     GRID POINT DENSITIES AT END OF STEP     O
24* CD    N       INTEGER           NUMBER OF INTERIOR GRID POINTS          I
25* CD    RBC     REAL              RIGHT BOUNDARY CONDITION FACTOR         I
26* CD    LBC     REAL              LEFT BOUNDARY CONDITION FACTOR          I
27* CD
28* CD    LANGUAGE AND LIMITATIONS:  THE SUBROUTINE ETBFCT IS A MULTIPLE-
29* CD    ENTRY FORTRAN ROUTINE IN SINGLE PRECISION (32 BITS ASC).  THE ASC
30* CD    PARAMETER STATEMENT IS USED TO SET SYMBOLICALLY THE INTERNAL ARRAY
31* CD    DIMENSIONS. UNDERFLOWS ARE POSSIBLE WHEN THE FUNCTION BEING TRANS-
32* CD    PORTED HAS MANY ZEROES. THE CALCULATIONS GENERALLY MISCONSERVE BY
33* CD    ONE OR TWO BITS PER CYCLE. THE RELATIVE PHASE AND AMPLITUDE ERRORS
34* CD    (FOR SMOOTH FUNCTIONS) ARE TYPICALLY SEVERAL PERCENT FOR CHARAC-
35* CD    TERISTIC LENGTHS OF 1 - 2 CELLS (WAVELENGTHS OF ORDER 10 CELLS).
36* CD    SHOCKS ARE GENERALLY ACCURATE TO BETTER THAN 1 PERCENT. THIS SUB-
37* CD    ROUTINE MUST BE COMPILED WITH THE Y OPTION TO FORCE STORAGE AND
38* CD    RETENTION OF INTERNAL VARIABLES. ALTERNATIVELY A COMMON BLOCK CAN
39* CD    BE ADDED TO ACCOMPLISH THE SAME END.
40* CD
41* CD    ENTRY POINTS: OGRIDE, NGRIDE, VELOCE, SOURCE, CONSRE,   SEE #11
42* CD    OF THE DETAILED DOCUMENTATION (OR THE LISTING BELOW) FOR THE EX-
43* CD    PLANATION AND USE OF THE ARGUMENTS TO THESE OTHER ENTRIES.
44* CD
45* CD    NO AUXILIARY OR LIBRARY ROUTINES ARE CALLED BY ETBFCT.
46* CD
47* CD    * * * * * * * * * * * * * * * * * * * * * * * * * * * * * * * * *
48* C
49*            PARAMETER  NPT = 202
50*            LOGICAL    LSOURC
51*            REAL       RBC, LBC, MASK1, MASK2, MASK3
```

```
52*            REAL       SOURCE(NPT), SCRH(NPT), RHOT(NPT), DIFF(NPT)
53*            REAL       ADUGTH(NPT), FLXH(NPT), NULH(NPT), MULH(NPT)
54*            REAL       LNRHOT(NPT), FSGN(NPT), FABS(NPT), EPSH(NPT)
55*            REAL       LORHOT(NPT), TERP(NPT), TERM(NPT), ADUDTH(NPT)
56*            REAL       LO(NPT), LN(NPT), LH(NPT), RLO(NPT), RLN(NPT)
57*            REAL       RNH(NPT), ROH(NPT), RLH(NPT), AH(NPT)
58*            DATA       MASK1, MASK2, MASK3 /Z80000000, 0.998, Z7FFFFFFF/
59*            DATA       ROH /NPT*1.0/, PI, FTPI /3.1415927, 4.1887902/
60*            DATA       SOURCE/NPT*0.0/, LSOURC/F/
61*            EQUIVALENCE (EPSH(1), SCRH(1)),        (LNRHOT(1), LORHOT(1))
62*            EQUIVALENCE (FLXH(1), SCRH(1)),        (FSGN(1), RHOT(1))
63*            EQUIVALENCE (FABS(1), SCRH(1)),        (TERP(1), SOURCE(1))
64*            EQUIVALENCE (TERM(1), SOURCE(1))
65* C
66* C     CALCULATE THE DIFFUSIVE AND CONVECTIVE FLUXES.
67*            NP = N + 1
68*            DO 11 I = 2, N
69*            FLXH(I) = 0.5*ADUDTH(I)*(RHOO(I) + RHOO(I-1))
70*      11    DIFF(I) = NULH(I)*(RHOO(I) - RHOO(I-1))
71*            RHOL = RHOO(1)*LBC
72*            RHOR = RHOO(N)*RBC
73*            DIFF(1) = NULH(1)*(RHOO(1) - RHOL)
74*            DIFF(NP) = NULH(NP)*(RHOR - RHOO(N))
75*            FLXH(1) = 0.5*ADUDTH(1)*(RHOO(1) + RHOL)
76*            FLXH(NP) = 0.5*ADUDTH(NP)*(RHOR + RHOO(N))
77* C
78* C     CALCULATE LAMBDAO*RHOT, THE TRANSPORTED MASS ELEMENTS.
79*            DO 12 I = 1, N
80*      12    LORHOT(I) = LO(I)*RHOO(I) - FLXH(I+1) + FLXH(I)
81* C
82* C     ADD IN THE SOURCE TERMS AS APPROPRIATE.
83*            IF (.NOT. LSOURC) GO TO 14
84*            DO 13 I = 1, N
85*      13    LORHOT(I) = LORHOT(I) + SOURCE(I)
86* C
87* C     CALCULATE THE PHOENICAL ANTIDIFFUSIVE FLUXES HERE.
88*      14    DO 16 I = 1, N
89*      16    RHOT(I) = LORHOT(I)*RLO(I)
90*            DO 17 I = 2, N
91*      17    FLXH(I) = MULH(I)*(RHOT(I) - RHOT(I-1))
92*            FLXH(1) = MULH(1)*(RHOT(1) - LBC*RHOT(1))
93*            FLXH(NP) = MULH(NP)*(RBC*RHOT(N) - RHOT(N))
94* C
95* C     DIFFUSE THE SOLUTION RHOT USING OLD FLUXES.
96*            DO 18 I = 1, N
97*      18    LNRHOT(I) = LORHOT(I) + DIFF(I+1) - DIFF(I)
98* C
99* C     CALCULATE THE TRANSPORTED/DIFFUSED DENSITY AND GRID DIFFERENCES.
100*           DO 19 I = 1, N
101*     19    RHOT(I) = LNRHOT(I)*RLN(I)
102*           DO 20 I = 2, N
103*     20    DIFF(I) = RHOT(I) - RHOT(I-1)
104*           DIFF(1) = RHOT(1) - LBC*RHOT(1)
```

```
105*              DIFF(NP) = RBC*RHOT(N) - RHOT(N)
106* C
107* C    CALCULATE THE SIGN AND MAGNITUDE OF THE ANTIDIFFUSIVE FLUX.
108*              DO 21 I = 1, NP
109*      21      FSGN(I) = AND(MASK1, DIFF(I))
110*              DO 22 I = 1, NP
111*              FABS(I) = AND(MASK3, FLXH(I))
112*      22      FSGN(I) = OR(MASK2, FSGN(I))
113* C
114* C    CALCULATE THE FLUX-LIMITING CHANGES ON THE RIGHT AND THE LEFT.
115*              DO 23 I = 1, N
116*      23      TERP(I) = FSGN(I)*LN(I)*DIFF(I+1)
117*              TERP(NP) = 1.0E+75
118*              DO 24 I = 1, NP
119*      24      FABS(I) = AMIN1(TERP(I), FABS(I))
120*              DO 25 I = 2, NP
121*      25      TERM(I) = FSGN(I)*LN(I-1)*DIFF(I-1)
122*              TERM(1) = 1.0E+75
123* C
124* C    CORRECT THE FLUXES COMPLETELY NOW.
125*              DO 35 I = 1, NP
126*      35      DIFF(I) = AMIN1(FABS(I), TERM(I))
127*              DO 36 I = 1, NP
128*      36      FLXH(I) = AMAX1(0.0, DIFF(I))
129*              DO 37 I = 1, NP
130*      37      FLXH(I) = FSGN(I)*FLXH(I)
131* C
132* C    CALCULATE THE NEW FLUX-CORRECTED DENSITIES.
133*              DO 40 I = 1, N
134*      40      LNRHOT(I) = LNRHOT(I) - FLXH(I+1) + FLXH(I)
135*              DO 41 I = 1, N
136*              SOURCE(I) = 0.0
137*      41      RHON(I) = LNRHOT(I)*RLN(I)
138*              LSOURC = .FALSE.
139*              RETURN
140* C
141* C
142* C    ----------------------------------------------------------------
143* C
144*              ENTRY VELOCE (U, N, UR, UL, DT)
145* C
146*              REAL     U(N)
147* C
148* CD    * * * * * * * * * * * * * * * * * * * * * * * * * * * * * * * *
149* CD
150* CD    VELOCE (U, N, UR, UL, DT)
151* CD    DESCRIPTION:  THIS ENTRY CALCULATES ALL VELOCITY-DEPENDANT COEFFS.
152* CD
153* CD    ARGUMENTS:
154* CD    U       REAL ARRAY(N)    FLOW VELOCITY AT THE GRID POINTS      I
155* CD    N       INTEGER          NUMBER OF INTERIOR GRID POINTS        I
156* CD    UR      REAL             VELOCITY OF FLOW AT RIGHT BOUNDARY    I
157* CD    UL      REAL             VELOCITY OF FLOW AT LEFT BOUNDARY     I
```

CIFER -- VERSION 06.03   DATE = 09/19/80   TIME = 09:59:44:92

```
158* CD    DT     REAL             STEPSIZE FOR THE TIME INTEGRATION    I
159* CD
160* CD    * * * * * * * * * * * * * * * * * * * * * * * * * * * * * * * * *
161* C
162* C      CALCULATE THE INTERFACE AREA X VELOCITY DIFFERENTIAL X DT.
163*              NP = N + 1
164*              DTH = 0.5*DT
165*              DO 101 I = 2, N
166*  101         ADUDTH(I) = AH(I)*DTH*(U(I) + U(I-1)) - ADUGTH(I)
167*              ADUDTH(1) = AH(1)*DT*UL - ADUGTH(1)
168*              ADUDTH(NP) = AH(NP)*DT*UR - ADUGTH(NP)
169* C
170* C      CALCULATE THE HALF-CELL EPSILON (V*DT/DX)
171*              DO 102 I = 1, NP
172*  102         EPSH(I) = ADUDTH(I)*RLH(I)
173* C
174* C      NEXT CALCULATE THE DIFFUSION AND ANTIDIFFUSION COEFFICIENTS.
175* C      VARIATION WITH EPSILON MEANS FOURTH-ORDER ACCURATE PHASES.
176*              DO 103 I = 1, NP
177*              NULH(I) = 0.167 + 0.333*EPSH(I)*EPSH(I)
178*  103         MULH(I) = 0.25 - 0.5*NULH(I)
179*              DO 104 I = 1, NP
180*              NULH(I) = LH(I)*NULH(I)
181*  104         MULH(I) = LH(I)*MULH(I)
182*              RETURN
183* C
184* C
185* C      ----------------------------------------------------------------
186* C
187*       ENTRY NGRIDE (RADN, N, RADR, RADL, ALPHA)
188* C
189*              INTEGER  ALPHA
190*              REAL     RADN(N)
191* C
192* CD    * * * * * * * * * * * * * * * * * * * * * * * * * * * * * * * * *
193* CD
194* CD    NGRIDE (RADN, N, RADR, RADL, ALPHA)
195* CD    DESCRIPTION:  THIS ENTRY SETS NEW GEOMETRY VARIABLES AND COEFFS.
196* CD
197* CD    ARGUMENTS:
198* CD    RADN    REAL ARRAY(N)      NEW GRID POINT POSITIONS            I
199* CD    N       INTEGER            NUMBER OF INTERIOR GRID POINTS      I
200* CD    RADR    REAL               POSITION OF THE RIGHT BOUNDARY      I
201* CD    RADL    REAL               POSITION OF THE LEFT BOUNDARY       I
202* CD    ALPHA   INTEGER            = 1    FOR CARTESIAN GEOMETRY       I
203* CD                               = 2    FOR CYLINDRICAL GEOMETRY     I
204* CD                               = 3    FOR SPHERICAL GEOMETRY       I
205* CD
206* CD    * * * * * * * * * * * * * * * * * * * * * * * * * * * * * * * * *
207* C
208* C      CALCULATE THE NEW HALF-CELL POSITIONS AND GRID CHANGES.
209*              NP = N + 1
210*              DO 202 I = 2, N
```

```
211*    202     RNH(I) = 0.5*(RADN(I) + RADN(I-1))
212*            RNH(1) = RADL
213*            RNH(NP) = RADR
214* C
215* C     CALCULATE THE THREE COORDINATE SYSTEMS.
216*            GO TO (203, 206, 209), ALPHA
217* C
218* C     CARTESIAN COORDINATES.
219*    203     DO 204 I = 1, NP
220*    204     AH(I) = 1.0
221*            DO 205 I = 1, N
222*    205     LN(I) = RNH(I+1) - RNH(I)
223*            GO TO 213
224* C
225* C     CYLINDRICAL COORDINATES.
226*    206     DO 207 I = 1, NP
227*            DIFF(I) = RNH(I)*RNH(I)
228*    207     AH(I) = PI*(ROH(I) + RNH(I))
229*            DO 208 I = 1, N
230*    208     LN(I) = PI*(DIFF(I+1) - DIFF(I))
231*            GO TO 213
232* C
233* C     SPHERICAL COORDINATES.
234*    209     DO 210 I = 1, NP
235*            DIFF(I) = RNH(I)*RNH(I)*RNH(I)
236*    210     SCRH(I) = (ROH(I) + RNH(I))*ROH(I)
237*            DO 211 I = 1, NP
238*    211     AH(I) = FTPI*(SCRH(I) + RNH(I)*RNH(I))
239*            DO 212 I = 1, N
240*    212     LN(I) = FTPI*(DIFF(I+1) - DIFF(I))
241* C
242* C     NOW THE GEOMETRIC VARIABLES WHICH ARE SYSTEM INDEPENDANT.
243*    213     DO 214 I = 2, N
244*    214     LH(I) = 0.5*(LN(I) + LN(I-1))
245*            LH(1) = LN(1)
246*            LH(NP) = LN(N)
247*            DO 215 I = 1, N
248*    215     RLN(I) = 1.0/LN(I)
249*            DO 216 I = 1, N
250*    216     ADUGTH(I) = AH(I)*(RNH(I) - ROH(I))
251*            DO 217 I = 2, N
252*    217     RLH(I) = 0.5*(RLN(I) + RLN(I-1))
253*            RLH(1) = RLN(1)
254*            RLH(NP) = RLN(N)
255*            RETURN
256* C
257* C
258* C     ----------------------------------------------------------------
259* C
260*        ENTRY SOURCE (N, DT, MODES, C, D, DR, DL)
261* C
262*            REAL    C(N), D(N)
263* C
```

```
264* CD    * * * * * * * * * * * * * * * * * * * * * * * * * * * * * * * * * * *
265* CD
266* CD    SOURCE (N, DT, MODES, C, D, DR, DL)
267* CD    DESCRIPTION:  THIS ENTRY ACCUMULATES DIFFERENT SOURCE TERMS.
268* CD
269* CD    ARGUMENTS:
270* CD    N       INTEGER          NUMBER OF INTERIOR GRID POINTS         I
271* CD    DT      REAL             STEPSIZE FOR THE TIME INTEGRATION      I
272* CD    MODES   INTEGER          = 1   COMPUTES + DIV (D)               I
273* CD                            = 2   COMPUTES + C*GRAD (D)            I
274* CD                            = 3   ADDS + D TO THE SOURCES          I
275* CD    C       REAL ARRAY(N)    ARRAY OF SOURCE VARIABLES AT GRID PTS  I
276* CD    D       REAL ARRAY(N)    ARRAY OF SOURCE VARIABLES AT GRID PTS  I
277* CD    DR      REAL             RIGHT BOUNDARY VALUE OF D              I
278* CD    DL      REAL             LEFT BOUNDARY VALUE OF D               I
279* CD
280* CD    * * * * * * * * * * * * * * * * * * * * * * * * * * * * * * * * * * *
281* C
282*           NP = N + 1
283*           DTH = 0.5*DT
284*           DTQ = 0.25*DT
285*           GO TO (310, 320, 330), MODES
286* C
287* C     + DIV(D) IS COMPUTED CONSERVATIVELY AND ADDED TO THE SOURCES.
288*    310    DO 311 I = 2, N
289*    311    SCRH(I) = DTH*AH(I)*(D(I) + D(I-1))
290*           SCRH(1) = DT*AH(1)*DL
291*           SCRH(NP) = DT*AH(NP)*DR
292*           DO 312 I = 1, N
293*    312    SOURCE(I) = SOURCE(I) + SCRH(I+1) - SCRH(I)
294*           LSOURC = .TRUE.
295*           RETURN
296* C
297* C     + C*GRAD(D) IS COMPUTED EFFICIENTLY AND ADDED TO THE SOURCES.
298*    320    DO 321 I = 2, N
299*    321    SCRH(I) = DTQ*(D(I) + D(I-1))
300*           SCRH(1) = DTH*DL
301*           SCRH(NP) = DTH*DR
302*           DO 322 I = 1, N
303*    322    DIFF(I) = SCRH(I+1) - SCRH(I)
304*           DO 323 I = 1, N
305*    323    SOURCE(I) = SOURCE(I) + C(I)*(AH(I+1) + AH(I))*DIFF(I)
306*           LSOURC = .TRUE.
307*           RETURN
308* C
309* C     + D IS ADDED TO THE SOURCES IN AN EXPLICIT FORMULATION.
310*    330    DO 331 I = 1, N
311*    331    SOURCE(I) = SOURCE(I) + DT*LO(I)*D(I)
312*           LSOURC = .TRUE.
313*           RETURN
314* C
315* C
316* C     ----------------------------------------------------------------
```

```
317* C
318*         ENTRY OGRIDE (N)
319* C
320* CD    * * * * * * * * * * * * * * * * * * * * * * * * * * * * * * * * * * * *
321* CD
322* CD    OGRIDE (N)
323* CD    DESCRIPTION:  THIS ENTRY COPIES OLD GRID AND GEOMETRY VARIABLES.
324* CD
325* CD    ARGUMENTS:
326* CD    N        INTEGER              NUMBER OF INTERIOR GRID POINTS        I
327* CD
328* CD    * * * * * * * * * * * * * * * * * * * * * * * * * * * * * * * * * * * *
329* C
330* C     COPY THE PREVIOUSLY NEW GRID VALUES TO BE USED AS THE OLD GRID.
331*              NP = N + 1
332*              DO 401 I = 1, N
333*              LO(I) = LN(I)
334*    401       RLO(I) = RLN(I)
335*              DO 402 I = 1, NP
336*    402       ROH(I) = RNH(I)
337*              RETURN
338* C
339* C
340* C     ----------------------------------------------------------------
341* C
342*         ENTRY CONSRE (RHO, N, CSUM)
343* C
344*              REAL     RHO(N)
345* C
346* CD    * * * * * * * * * * * * * * * * * * * * * * * * * * * * * * * * * * * *
347* CD
348* CD    CONSRE (RHO, N, CSUM)
349* CD    DESCRIPTION:  THIS ENTRY COMPUTES THE OSTENSIBLY CONSERVED SUM.
350* CD
351* CD    ARGUMENTS:
352* CD    RHO      REAL ARRAY(N)       GRID PT VALUES FOR CONSERVATION SUM    I
353* CD    N        INTEGER             NUMBER OF INTERIOR GRID POINTS         I
354* CD    CSUM     REAL                VALUE OF THE CONSERVATION SUM OF RHO   O
355* CD
356* CD    * * * * * * * * * * * * * * * * * * * * * * * * * * * * * * * * * * * *
357* CD
358* C     COMPUTE THE OSTENSIBLY CONSERVED TOTAL MASS (BEWARE YOUR B.C.)
359*              CSUM = 0.0
360*              DO 501 I = 1, N
361*    501       CSUM = CSUM + LN(I)*RHO(I)
362*              RETURN
363*         END
```

Appendix B

```
 1* C
 2* C
 3* C        FLIMIT IS A VECTORIZED ASC FORTRAN MODULE WHICH IMPLEMENTS ZALESAK'S
 4* C        MULTIDIMENSIONAL FLUX LIMITER [JCP 31, 335 (1979)] IN 2D CARTESIAN
 5* C        GEOMETRY.  IT INCORPORATES AS OPTIONS A "PRE-LIMITING" STEP UTILIZING
 6* C        THE STRONG 1-D LIMITER OF BORIS AND BOOK (WHEN JPRLIM = .TRUE. ),
 7* C        AND THE ABILITY TO LOOK BACK TO THE PREVIOUS TIMESTEP FOR UPPER AND
 8* C        LOWER BOUNDS ON THE NEW SOLUTION ( WHEN FOLD = .TRUE. ).
 9* C
10* C
11* C                             UTILIZATION
12* C
13* C        FLIMIT          - SUBROUTINE CALLED AT EVERY TIMESTEP FOR EACH
14* C                           CONVECTIVE EQUATION BEING SOLVED
15* C
16* C        AUXILIARY ROUTINES CALLED BY ABOVE:
17* C
18* C        GBCOND          - APPLY SIMPLE PERIODIC OR REFLECTING BOUNDARY
19* C                           CONDITIONS TO SCRATCH AND INTERMEDIATE ARRAYS
20* C                           AFTER INTERIOR VALUES ARE KNOWN
21* C
22* C                             CALLING SEQUENCE
23* C
24* C
25* C        FLX, FLY        - RAW (UNLIMITED) ANTIDIFFUSIVE FLUXES. DIMENSIONALLY
26* C                           THESE FLUXES SHOULD BE IN THE SAME UNITS AS
27* C                           FTD (BELOW) MULTIPLIED BY AN AREA
28* C
29* C        FTD             - ARRAY CONTAINING THE TIME-ADVANCED, LOW ORDER
30* C                           ("TRANSPORTED AND DIFFUSED") SOLUTION
31* C
32* C        NX, NY          - DIMENSIONS OF MESH (RESTRICTED AT COMPILE TIME
33* C                           BY VARIABLE DIMENSIONING TO NX <= MX AND NY <= MY,
34* C                           WHERE MX AND MY ARE DEFINED IN PARAMETER STATEMENT
35* C
36* C        SX, SY, SM      - SCRATCH ARRAYS
37* C
38* C        DX, DY          - ARRAYS OF CELL SIZES IN X, Y DIRECTIONS
39* C                           THESE QUANTITIES ARE TAKEN TO BE "CELL CENTERED".
40* C                           THAT IS, THE AREA OF THE CELL CENTERED ON GRID POINT
41* C                           (I,J) IS DX(I)*DY(J) .
42* C
43* C        TX, TY          - CORRECTED ANTIDIFFUSIVE FLUXES RETURNED BY FLIMIT
44* C
45* C        FAA             - ARRAY CONTAINING THE SOLUTION FROM THE PREVIOUS
46* C                           TIMESTEP (USED ONLY IF FOLD = .TRUE. )
47* C
48* C        RBCX, RBCY      - BOUNDARY CONDITION PARAMETERS ( SEE COMMENTS IN
49* C                           SUBROUTINE GBCOND )
50* C
51* C
```

```
52* C       THE COMMON BLOCK /LIMIT/ CONTAINS THREE SCALAR LOGICAL VARIABLES
53* C       WHICH MAY BE SET BY THE USER FROM OUTSIDE THE SUBROUTINE:
54* C
55* C       PRLIM                    SETTING PRLIM = .TRUE. ZEROES THE FLUX WHENEVER
56* C                                ITS SIGN DIFFERS FROM THAT OF THE CORRESPONDING
57* C                                FIRST DIFFERENCE IN FTD - IT IS A SIMPLIFIED VERSION
58* C                                OF EQ (14') IN JCP 31, PG 349.
59* C
60* C       JPRLIM                   SETTING JPRLIM = .TRUE. CAUSES THE FLUXES TO BE
61* C                                "PRE-LIMITED" USING THE STRONG 1-D LIMITER OF BORIS
62* C                                AND BOOK BEFORE PASSING THE RESIDUAL FLUXES ON TO
63* C                                THE MULTIDIMENSIONAL LIMITER - SEE JCP 31, PP 349-350
64* C
65* C       FOLD                     SETTING FOLD = .TRUE. ALLOWS THE LIMITER TO LOOK BACK
66* C                                TO THE SOLUTION FROM THE PREVIOUS TIMESTEP (WHICH
67* C                                MUST BE STORED IN ARRAY FAA ) TO FIND UPPER AND
68* C                                LOWER BOUNDS ON THE NEW SOLUTION
69* C
70* C       THE MOST CONSERVATIVE (I.E., MOST DIFFUSIVE) CHOICE IS OBTAINED BY SETTING
71* C       JPRLIM = .TRUE. AND FOLD = .FALSE. ( PRLIM IS IRRELEVANT WHEN JPRLIM
72* C       IS .TRUE. )  THIS IS THE RECOMMENDED CHOICE FOR THE FIRST ATTEMPT.
73* C
74* C       THE BOUNDARY CONDITION SUBROUTINE GBCOND CAN BE REPLACED BY A MORE
75* C       COMPLICATED PROBLEM-DEPENDENT PRESCRIPTION WHICH VARIES AS A FUNCTION
76* C       OF THE ARRAY BEING BOUNDED.
77* C
78* C
79*         SUBROUTINE FLIMIT ( FLX, FLY, FTD, NX, NY, SX, SY, DX, DY,
80*    1            TX, TY, SM, FAA, RBCX, RBCY )
81*         PARAMETER MX=98,  MY=98,  MXNX=260
82*         DIMENSION FLX(NX/MX/,NY/MY/), FLY(NX/MX/,NY/MY/),
83*    1            FTD(NX/MX/,NY/MY/), SX(NX/MX/,NY/MY/), SY(NX/MX/,NY/MY/),
84*    2            TX(NX/MX/,NY/MY/), TY(NX/MX/,NY/MY/), SM(NX/MX/,NY/MY/)
85*         DIMENSION FAA(NX/MX/,NY/MY/)
86*         DIMENSION DX(NX/MX/), DY(NY/MY/)
87*         DIMENSION FTDO(MXNX), FTDP(MXNX)
88*         COMMON /LIMIT/ PRLIM, JPRLIM, FOLD
89*         LOGICAL PRLIM, JPRLIM, FOLD
90* C       NOTE THAT FLX AND FLY MUST BE REAL FLUXES (LIKE GRAMS FOR EX.)
91* C       FLX(I,J) IS CENTERED BETWEEN FTD(I,J) AND FTD(I+1,J)
92* C       FLY(I,J) IS CENTERED BETWEEN FTD(I,J) AND FTD(I,J+1)
93* C       FLX(I,J) DEFINED I=1,NXM  J=1,NY
94* C       FLY(I,J) DEFINED J=1,NYM  I=1,NX
95*         NXM=NX-1
96*         NYM=NY-1
97*         PBCX = 1.0 - RBCX**2
98*         PBCY = 1.0 - RBCY**2
99*         ARBCX = ABS(RBCX)
100*        ARBCY = ABS(RBCY)
101*        IF ( .NOT. JPRLIM ) GO TO 64
102*        DO 52 J=1,NY
103*        DO 52 I=2,NXM
104*        SX(I,J) = (FTD(I+1,J)-FTD(I,J))*DX(I)*DY(J)
```

```
105*     52      SY(I,J) = (FTD(I,J)-FTD(I-1,J))*DX(I)*DY(J)
106*             DO 521 J=1,NY
107*             FTDO(J) = PBCX*FTD(NX-2,J) + RBCX*FTD(3,J)
108*             FTDP(J) = RBCX*FTD(NX-2,J) + PBCX*FTD(3,J)
109*             SY(1,J) = (FTD(1,J)-FTDO(J))*DX(1)*DY(J)
110*    521      SX(NX,J) = (FTDP(J)-FTD(NX,J))*DX(NX)*DY(J)
111*             DO 53 J=1,NY
112*             DO 53 I=1,NXM
113*             SM(I,J) = SIGN(1.0,FLX(I,J))
114*             FLX(I,J) = ABS(FLX(I,J))
115*             TY(I,J) = SM(I,J)*SY(I,J)
116*             FLX(I,J) = AMIN1( FLX(I,J), TY(I,J) )
117*             TY(I,J) = SM(I,J)*SX(I+1,J)
118*             FLX(I,J) = AMIN1( FLX(I,J), TY(I,J) )
119*             FLX(I,J) = AMAX1( FLX(I,J), 0.0 )
120*     53      FLX(I,J) = FLX(I,J)*SM(I,J)
121*             DO 62 J=2,NYM
122*             DO 62 I=1,NX
123*             SX(I,J) = (FTD(I,J+1)-FTD(I,J))*DX(I)*DY(J)
124*     62      SY(I,J) = (FTD(I,J)-FTD(I,J-1))*DX(I)*DY(J)
125*             DO 621 I=1,NX
126*             FTDO(I) = PBCY*FTD(I,NY-2) + RBCY*FTD(I,3)
127*             FTDP(I) = RBCY*FTD(I,NY-2) + PBCY*FTD(I,3)
128*             SY(I,1) = (FTD(I,1)-FTDO(I))*DX(I)*DY(1)
129*    621      SX(I,NY) = (FTDP(I)-FTD(I,NY))*DX(I)*DY(NY)
130*             DO 63 J=1,NYM
131*             DO 63 I=1,NX
132*             SM(I,J) = SIGN(1.0,FLY(I,J))
133*             FLY(I,J) = ABS(FLY(I,J))
134*             TY(I,J) = SM(I,J)*SY(I,J)
135*             FLY(I,J) = AMIN1( FLY(I,J), TY(I,J) )
136*             TY(I,J) = SM(I,J)*SX(I,J+1)
137*             FLY(I,J) = AMIN1( FLY(I,J), TY(I,J) )
138*             FLY(I,J) = AMAX1( FLY(I,J), 0.0 )
139*     63      FLY(I,J) = FLY(I,J)*SM(I,J)
140*     64      CONTINUE
141*             IF ( .NOT. PRLIM ) GO TO 5
142*             DO 2 J=1,NY
143*             DO 2 I=1,NXM
144*             SM(I,J) = FLX(I,J)*(FTD(I+1,J)-FTD(I,J))
145*      2      FLX(I,J) = FLX(I,J)*AMAX1(0.0,SIGN(1.0,SM(I,J)) )
146*             DO 4 J=1,NYM
147*             DO 4 I=1,NX
148*             SM(I,J) = FLY(I,J)*(FTD(I,J+1)-FTD(I,J))
149*      4      FLY(I,J) = FLY(I,J)*AMAX1(0.0,SIGN(1.0,SM(I,J)) )
150*      5      CONTINUE
151*             DO 100 J=1,NY
152*             DO 100 I=1,NX
153*             SX(I,J) = 1.0
154*             SY(I,J) = 1.0
155*    100      CONTINUE
156*             IF ( FOLD ) GO TO 203
157*             DO 202 J=1,NY
```

```
158*              DO 202 I=1,NX
159*     202      SM(I,J) = FTD(I,J)
160*              GO TO 2041
161*     203      DO 204 J=1,NY
162*              DO 204 I=1,NX
163*     204      SM(I,J) = AMAX1(FTD(I,J),FAA(I,J))
164*     2041     DO 205 J=2,NYM
165*              DO 205 I=2,NXM
166*              TX(I,J)=(AMAX1( SM(I-1,J), SM(I,J), SM(I+1,J), SM(I,J-1),
167*       1            SM(I,J+1)) - FTD(I,J) )*DX(I)*DY(J)
168*     205      CONTINUE
169*              DO 206 J=2,NYM
170*              DO 206 I=2,NXM
171*              SM(I,J) = AMAX1(0.0,FLX(I-1,J)) - AMIN1(0.0,FLX(I,J))
172*     206      CONTINUE
173*              DO 210 J=2,NYM
174*              DO 210 I=2,NXM
175*              SM (I,J) = SM(I,J)
176*       1          + ( AMAX1(0.0,FLY(I,J-1)) - AMIN1(0.0,FLY(I,J)) )
177*              TY(I,J) = 1.0
178*     210      CONTINUE
179*              DO 230 J=2,NYM
180*              DO 230 I=2,NXM
181*              IF ( SM(I,J) .LE. TX(I,J) ) GO TO 230
182*              TY(I,J) = TX(I,J)/SM(I,J)
183*     230      CONTINUE
184*              DO 240 J=2,NYM
185*              DO 240 I=2,NXM
186*              TX(I,J) = TY(I,J)
187*     240      CONTINUE
188*              CALL GBCOND ( NX, NY, TX, ARBCX, ARBCY )
189*              DO 250 J=1,NY
190*              DO 250 I=1,NXM
191*              SM(I,J) = -SIGN(TX(I,J),FLX(I,J))
192*     250      CONTINUE
193*              DO 251 J=1,NY
194*              DO 251 I=1,NXM
195*              SM(I,J) = AMAX1( SM(I,J),  SIGN(TX(I+1,J),FLX(I,J)) )
196*     251      CONTINUE
197*              DO 260 J=1,NY
198*              DO 260 I=1,NXM
199*              SX(I,J) = AMIN1(SX(I,J),SM(I,J))
200*     260      CONTINUE
201*              DO 270 J=1,NYM
202*              DO 270 I=1,NX
203*              SM(I,J) = -SIGN(TX(I,J),FLY(I,J))
204*     270      CONTINUE
205*              DO 271 J=1,NYM
206*              DO 271 I=1,NX
207*              SM(I,J) = AMAX1( SM(I,J),  SIGN(TX(I,J+1),FLY(I,J)) )
208*     271      CONTINUE
209*              DO 280 J=1,NYM
210*              DO 280 I=1,NX
```

```
211*            SY(I,J) = AMIN1(SY(I,J),SM(I,J))
212*    280     CONTINUE
213*            IF ( FOLD ) GO TO 403
214*            DO 400 J=1,NY
215*            DO 400 I=1,NX
216*    400     SM(I,J) = FTD(I,J)
217*            GO TO 4041
218*    403     DO 404 J=1,NY
219*            DO 404 I=1,NX
220*    404     SM(I,J) = AMIN1(FTD(I,J),FAA(I,J))
221*   4041     DO 405 J=2,NYM
222*            DO 405 I=2,NXM
223*            TX(I,J)=(FTD(I,J)-AMIN1( SM(I-1,J), SM(I,J), SM(I+1,J),
224*      1          SM(I,J-1), SM(I,J+1)) )*DX(I)*DY(J)
225*    405     CONTINUE
226*            DO 401 J=1,NY
227*            DO 401 I=1,NXM
228*    401     TY(I,J) = SX(I,J)*FLX(I,J)
229*            DO 406 J=2,NYM
230*            DO 406 I=2,NXM
231*            SM(I,J) = AMAX1(0.0, TY(I,J)) - AMIN1(0.0,TY(I-1,J))
232*    406     CONTINUE
233*            DO 402 J=1,NYM
234*            DO 402 I=1,NX
235*    402     TY(I,J) = SY(I,J)*FLY(I,J)
236*            DO 410 J=2,NYM
237*            DO 410 I=2,NXM
238*            SM (I,J) = SM(I,J)
239*      1          +(AMAX1(0.0, TY(I,J)) - AMIN1(0.0, TY(I,J-1)) )
240*            TY(I,J) = 1.0
241*    410     CONTINUE
242*            DO 430 J=2,NYM
243*            DO 430 I=2,NXM
244*            IF ( SM(I,J) .LE. TX(I,J) ) GO TO 430
245*            TY(I,J) = TX(I,J)/SM(I,J)
246*    430     CONTINUE
247*            DO 440 J=2,NYM
248*            DO 440 I=2,NXM
249*            TX(I,J) = TY(I,J)
250*    440     CONTINUE
251*            CALL GBCOND ( NX, NY, TX, ARBCX, ARBCY )
252*            DO 450 J=1,NY
253*            DO 450 I=1,NXM
254*            SM(I,J) = SIGN(TX(I,J),FLX(I,J))
255*    450     CONTINUE
256*            DO 451 J=1,NY
257*            DO 451 I=1,NXM
258*            SM(I,J) = AMAX1( SM(I,J),-SIGN(TX(I+1,J),FLX(I,J)) )
259*    451     CONTINUE
260*            DO 460 J=1,NY
261*            DO 460 I=1,NXM
262*            SX(I,J) = SX(I,J)*SM(I,J)
263*    460     CONTINUE
```

```
264*              DO 470 J=1,NYM
265*              DO 470 I=1,NX
266*              SM(I,J) = SIGN(TX(I,J),FLY(I,J))
267*     470      CONTINUE
268*              DO 471 J=1,NYM
269*              DO 471 I=1,NX
270*              SM(I,J) = AMAX1( SM(I,J), -SIGN(TX(I,J+1),FLY(I,J)) )
271*     471      CONTINUE
272*              DO 480 J=1,NYM
273*              DO 480 I=1,NX
274*              SY(I,J) = SY(I,J)*SM(I,J)
275*     480      CONTINUE
276*              DO 500 J=1,NY
277*              DO 500 I=1,NXM
278*              TX (I,J) = FLX(I,J)*SX(I,J)
279*     500      CONTINUE
280*              DO 600 J=1,NYM
281*              DO 600 I=1,NX
282*              TY (I,J) = FLY(I,J)*SY(I,J)
283*     600      CONTINUE
284*              RETURN
285*              END
286* C
287* C
288*              SUBROUTINE GBCOND ( NX, NY, F, RBCX, RBCY )
289* C
290* C        GBCOND IS A VERY SIMPLE ROUTINE FOR SETTING BOUNDARY CONDITIONS
291* C        ON THE NX BY NY ARRAY F.
292* C        RBCX (FOR THE X-DIRECTION) AND RBCY (FOR THE Y-DIRECTION) ARE REAL
293* C        SCALARS WHICH MAY TAKE ON VALUES -1.0, 0.0, AND 1.0:
294* C             RBC = -1.0 - ANTISYMMETRIC BOUNDARY CONDITIONS
295* C             RBC =  0.0 - PERIODIC BOUNDARY CONDITIONS
296* C             RBC =  1.0 - SYMMETRIC BOUNDARY CONDITIONS
297* C
298*              PARAMETER MX=98, MY=98
299*              DIMENSION F(NX/MX/,NY/MY/)
300*              NXM = NX-1
301*              NYM = NY-1
302*              PBCX = 1.0 - RBCX**2
303*              PBCY = 1.0 - RBCY**2
304*              DO 1 J=2,NYM
305*              F(1,J) = PBCX*F(NXM,J) + RBCX*F(2,J)
306*     1        F(NX,J)= RBCX*F(NXM,J) + PBCX*F(2,J)
307*              DO 2 I=1,NX
308*              F(I,1) = PBCY*F(I,NYM) + RBCY*F(I,2)
309*     2        F(I,NY)= RBCY*F(I,NYM) + PBCY*F(I,2)
310*              RETURN
311*              END
```

Appendix C

```
  1*        SUBROUTINE ADINC (RAD, VEL, RHO, PRE, N, DTIN, CYCLE,
  2*      1                   RRNEW, RLNEW, VRNEW, VLNEW)
  3* C
  4* C
  5* C        ADINC HAS BEEN CONSTRUCTED AS A UTILITY PACKAGE TO ADVANCE THE
  6* C    FOUR HYDRODYNAMIC VARIABLES..
  7* C
  8* C    RAD(I)       = POSITION (RADIUS) OF THE I-TH CELL INTERFACE (CM)
  9* C    VEL(I)       = VELOCITY OF THE I-TH CELL INTERFACE (CM/SEC)
 10* C    RHO(I)       = DENSITY IN CELL I BETWEEN INTERFACES I,I+1 (GM/CC)
 11* C    PRE(I)       = PRESSURE IN THE I-TH COMPUTATIONAL CELL (ERG/CC)
 12* C
 13* C    LAGRANGIAN FLUID DYNAMICS EQUATIONS ARE SOLVED INCLUDING A FLEX-
 14* C    IBLE EQUATION OF STATE WHICH CAN VARY FROM CELL TO CELL IN THE
 15* C    DISCRETIZED REPRESENTATION OF THE FLUID. THE EQUATIONS SOLVED ARE
 16* C
 17* C        D(RAD)              D(VEL)     - 1
 18* C        ------  = VEL,      ------  = ----- GRAD (PRE),
 19* C          DT                  DT       RHO
 20* C
 21* C    AND THE EQUATION OF STATE ..
 22* C
 23* C                        ( PRE ) ** 1/GAMMAC
 24* C        RHO  =  RHOC +  (-----)
 25* C                        ( ENTC).
 26* C
 27* C    ALL NON-IDEAL EFFECTS WHICH MIGHT BE INCLUDED IN AN ADINC CALCU-
 28* C    LATION HAVE TO BE INCLUDED SEPARATELY EITHER BY PHENOMENOLOGICALLY
 29* C    IMBEDDING A SIMPLE MODEL IN THE CALCULATION OR BY TIMESTEP
 30* C    SPLITTING.
 31* C
 32* C        EACH FULLY LAGRANGIAN CELL HAS SEVERAL QUANTITIES THAT ARE
 33* C    CONSERVED MOVING WITH THE FLUID AS LONG AS DIFFUSIVE AND OTHER
 34* C    NON-IDEAL EFFECTS AND SOURCE TERMS ARE NOT INCLUDED IN THE CALCU-
 35* C    LATION. THE EQUATION OF STATE IN EACH FLUID CELL MAY DIFFER.
 36* C    THE QUANTITIES INVOLVED IN THE EQUATION OF STATE ARE INITIALIZED
 37* C    BY THE TWO ENTRIES SETMAT AND SETEOS AND THE EQUATION OF STATE IS
 38* C    EVALUATED BY CALLING USEEOS. THE EQUATION OF STATE QUANTITIES
 39* C    ARE COMMUNICATED THROUGHOUT THE ADINC PACKAGE IN COMMON BLOCK
 40* C    /ADICOM/. THESE "CONSTANTS" VARY FROM CELL TO CELL ACCORDING TO
 41* C    THE INITIAL CONDITIONS. FOLLOWING ARE THE DEFINITIONS OF THESE
 42* C    QUANTITIES..
 43* C
 44* C    MATERC(I)    = CELL IDENTIFIER = L.MM WHERE 0 < L < 10 IS THE
 45* C                           LAYER NUMBER AND 0 < MM < 100 IS
 46* C                           THE MATERIAL IDENTIFIER.
 47* C    MASSC(I)     = CELL MASS = RHO(I)*LAM(I) - HELD CONSTANT IN ADINC
 48* C    GAMMAC(I)    = CELL ADIABATIC GAS CONSTANT - HELD FIXED IN ADINC
 49* C    ENTC(I)      = CELL ENTROPY - CONSTANT DURING ADINC HYDRODYNAMICS
 50* C    RHOC(I)      = DENSITY CONSTANT IN THE EQUATION OF STATE (GM/CC)
 51* C    NCELLS       = NUMBER OF CELLS OF FLUID IN THE CALCULATION
```

```
52* C
53* C          VARIABLE GAMMA AND ENTROPY ARE USED IN EACH LAGRANGIAN CELL.
54* C     AN IMPLICIT PRESSURE ITERATION ENSURES LINEAR STABILITY BUT HIGH
55* C     FREQUENCY PHENOMENA ARE INACCURATELY INTEGRATED WHEN THE TIMESTEPS
56* C     ARE CHOSEN TO BE APPRECIABLY LONGER THAN THE COURANT TIMESTEP. THE
57* C     NONLINEAR TERMS ARE ITERATED WITH A QUADRATICALLY CONVERGENT ALGO-
58* C     RITHM. REFERENCE: NRL MEMORANDUM REPORT #4022, 1979.
59* C
60* C          PROBLEMS IN ONE OF FOUR GEOMETRIES CAN BE SET UP FOR ADINC BY
61* C     CHANGING THE INTEGER ALPHA IN THE CALL TO SETGEO...
62* C
63* C     ALPHA = 1    CARTESIAN COORDINATES.
64* C     ALPHA = 2    CYLINDRICAL COORDINATES.
65* C     ALPHA = 3    SPHERICAL COORDINATES.
66* C     ALPHA = 4    POWER SERIES COORDINATES.
67* C
68* C     ADINC USES THE UTILITY USEGEO TO DETERMINE THE INSTANTANEOUS GRID
69* C     QUANTITIES. CALL USEGEO CALCULATES THE INTERFACE AREAS, THE CELL
70* C     VOLUMES, AND THE CELL CENTER POSITIONS. ADINC DOES NOT UPDATE ALL
71* C     THE GEOMETRY QUANTITIES AUTOMATICALLY ON EXIT. THUS USEGEO MUST
72* C     ALSO BE CALLED EXTERNALLY. IN GENERAL ADINC DOES NOT HAVE ACCESS
73* C     TO THE USER-DEFINED ARRAYS FOR AREA, RADC, AND LAMC.
74* C
75* C          THE BOUNDARY CONDITIONS TREATED BY ADINC ARE QUITE GENERAL.
76* C     THE POSITIONS AND VELOCITIES OF THE REGION BOUNDING INTERFACES
77* C     RAD(1) AND RAD(N+1) CAN BE EXTERNALLY DETERMINED FUNCTIONS OF TIME
78* C     AND OTHER PHYSICAL VARIABLES DURING THE CALCULATION. THE BOUNDARY
79* C     CONDITIONS ARE COMMUNICATED TO THE ADINC PACKAGE VIA THE FOUR
80* C     ARGUMENTS...
81* C
82* C     RRNEW      = RIGHT BOUNDARY POSITION RAD(N+1) AT END OF TIMESTEP
83* C     RLNEW      = LEFT BOUNDARY POSITION RAD(1) AT END OF TIMESTEP
84* C     VRNEW      = RIGHT BOUNDARY VELOCITY VEL(N+1) AT END OF TIMESTEP
85* C     VLNEW      = LEFT BOUNDARY VELOCITY VEL(1) AT END OF TIMESTEP
86* C
87* C          SEVERAL AUXILIARY VARIABLES ARE USED BY ADINC ITSELF OR THE
88* C     ADINC ROUTINES WHICH SHOULD ALSO BE EXPLAINED TO THE USER.
89* C
90* C     N = NCELLS = NUMBER OF FLUID CELLS IN THE ADINC INTEGRATION
91* C     DTIN       = TIME INTERVAL FOR THE ADINC INTEGRATION WHICH MAY
92* C                  SUBCYCLE UP TO 100 TIMES INTERNALLY IF NEEDED FOR
93* C                  ACCURACY OR STABILITY.
94* C     CYCLE      = TIMESTEP NUMBER USED BY ADINC FOR IDENTIFICATION
95* C     EPSRO      = EXPLICITNESS PARAMETER FOR THE POSITION INTEGRATION
96* C     EPSVO      = EXPLICITNESS PARAMETER FOR THE VELOCITY INTEGRATION
97* C     NDAMP      = NUMBER OF CYCLES AT THE BEGINNING OF A CALCULATION
98* C                  IN WHICH ADDITIONAL DAMPING/SMOOTHING IS APPLIED.
99* C     EPSR       = EXPLICITNESS PARAMETER FOR RAD LAST USED BY ADINC
100* C    EPSV       = EXPLICITNESS PARAMETER FOR VEL LAST USED BY ADINC
101* C
102* C          THE PARAMETER NPT, HERE 202, MUST BE AT LEAST TWO LARGER THAN
103* C     THE NUMBER OF FINITE DIFFERENCE CELLS BEING INTEGRATED BY ADINC.
104* C
```

```
105*  C
106*        PARAMETER     NPT = 202, MPT = 2*NPT
107*        INTEGER   CYCLE
108*        REAL*8    RAD(NPT),    VEL(NPT),    RHO(NPT),    PRE(NPT)
109*        REAL*8    RHOO(NPT),   RHON(NPT),   PREO(NPT),   PREN(NPT)
110*        REAL*8    LAMO(NPT),   LAMN(NPT),   RADCO(NPT),  RADCN(NPT)
111*        REAL*8    RIN(NPT),    VIN(NPT),    DLAM(NPT),   LAMEOS(NPT)
112*        REAL*8    RBARI(NPT),  RBAR(NPT),   RDRI(NPT),   DTORHO(NPT)
113*        REAL*8    AA(MPT),     BB(MPT),     CC(MPT),     DD(MPT)
114*        REAL*8    AI(NPT),     BI(NPT),     C(NPT),      DP(NPT)
115*        REAL*8    DM(NPT),     SCA(MPT),    SCB(MPT),    DLAMDP(NPT)
116*        REAL*8    AREAO(NPT),  AREAH(NPT),  AREAN(NPT)
117*        REAL*8    ERRLIM,      ERRMAX,      ERROR,       AMNMAS
118*        REAL*8    PREMAX,      PREMIN,      RPREMX,      DT
119*        REAL*8    VRNEW,       VLNEW,       RRNEW,       RLNEW
120*        REAL*8    VROLD,       VLOLD,       RROLD,       RLOLD
121*        REAL*8    DVROLD,      DVLOLD,      DRROLD,      DRLOLD
122*        REAL*8    EPSR,        OMEPSR,      EPSV,        OMEPSV
123*        REAL*8    EPSRO,       EPSVO,       OMERDT,      AMXMAS
124*        REAL*8    DTIN,        DTVAL,       FGOHED
125*  C
126*  C     DECLARATIONS FOR COMMON BLOCK /ADICOM/ APPEAR THROUGHOUT THE ADINC
127*  C     PACKAGE ROUTINES.
128*        REAL*8    MASSC(NPT), ENTC(NPT), RHOC(NPT), MATERC(NPT)
129*        REAL*8    GAMMAC(NPT)
130*        COMMON    /ADICOM/ MATERC, MASSC, GAMMAC, ENTC, RHOC, NCELLS
131*  C
132*        DATA      NDAMP /10/,    EPSR, EPSV /0.0D0, 0.0D0/
133*        DATA      NCALL, NITER, NTIME /0, 0, 0/, FGOHED /0.01D0/
134*        DATA      ERRLIM, EPSRO,EPSVO,ITEMAX/1.0D-9, 0.45D0, 0.45D0, 6/
135*        EQUIVALENCE  (DTORHO(1), RDRI(1)),      (AREAH(1), AREAN(1))
136*  C
137*  C     ADINC FORMATS FOR DIAGNOSTIC AND ERROR PRINTS.
138*  1001  FORMAT ('OADINC TIMESTEP PROBLEM AT CYCLE', I5, ' DT = ',
139*     1       1PD12.4, ' AND ITEMAX = ', I2, ' ERRLIM = ', D12.4, /,
140*     2       10X, ' ERRMAX AT', I4, ' = ', D12.4, ' AND', I3,
141*     3       ' CELLS NOT CONVERGED.')
142*  1002  FORMAT (' ADINC TIMESTEP PROBLEM AT CYCLE', I5, ' DTVAL = ',
143*     1       1PD12.4, ' BUT DTIN = ', D12.4, ' TIMESTEP SUBCYCLED.')
144*  1003  FORMAT ('OADINC TIMESTEP PROBLEM AT CYCLE', I5, ' DTVAL = ',
145*     1       1PD12.4, ' BUT DTIN = ', D12.4, ' CALCULATION STOPPED.')
146*  1004  FORMAT ('OADINC INPUT PROBLEM AT CYCLE', I5,
147*     1       ' THE SYSTEM SIZE N ', I4, ' OUT OF RANGE. NPT =', I4,
148*     2       ' CALCULATION STOPPED.')
149*  1005  FORMAT ('OADINC INPUT PROBLEM AT CYCLE', I5,
150*     1       ' THE LEFT BOUNDARY HAS CROSSED THE RIGHT .', 1P2D12.4,
151*     2       ' CALCULATION STOPPED.')
152*  1006  FORMAT ('OADINC INPUT PROBLEM AT CYCLE', I5,
153*     1       ' THE CELL SIZE WAS ', 1PD12.4,' AT CELL ', I4,
154*     2       ' CALCULATION STOPPED.')
155*  1007  FORMAT ('OADINC INPUT PROBLEM AT CYCLE', I5,
156*     1       ' THE DENSITY MIN WAS ', 1PD12.4,' AT CELL ', I4,
157*     2       ' CALCULATION STOPPED.')
```

```
158*  1008      FORMAT ('0ADINC INPUT PROBLEM AT CYCLE', I5,
159*     1              ' THE PRESSURE MIN WAS ', 1PD12.4,' AT CELL ', I4,
160*     2              ' CALCULATION STOPPED.')
161*  1009      FORMAT ('0ADINC FREQUENCY COUNTERS (SINCE LAST CHECK) AT ',
162*     1              'CYCLE', I5, /, 10X, 'NO. CALLS =', I5, '    NO. TIME',
163*     2              'STEPS =', I5, '     TOTAL NO. ITERATIONS =', I5, /)
164* C
165* C
166* C       CHECK THE INPUT TO ADINC FOR REASONABLENESS.
167*             NPTO = NPT
168*             IF ( N.LE.2 .OR. N.GT.NPT-2) WRITE (6, 1004) CYCLE, N, NPTO
169*             IF ( N.LE.2 .OR. N.GT.NPT-2) STOP
170*             IF (RRNEW.LE.RLNEW) WRITE (6, 1005) CYCLE, RLNEW, RRNEW
171*             IF (RRNEW.LE.RLNEW) STOP
172*             DO 20 I = 1, N
173*      20      DM(I+1) = RAD(I+1) - RAD(I)
174*             CALL MAXMIN (DM(2), N, PREMAX, IMAX, PREMIN, IMIN)
175*             IF (PREMIN .LE. 0.0D0) WRITE (6, 1006) CYCLE, PREMIN, IMIN
176*             IF (PREMIN .LE. 0.0D0)  STOP
177*             CALL MAXMIN (RHO(2), N, PREMAX, IMAX, PREMIN, IMIN)
178*             IF (PREMIN .LE. 0.0D0) WRITE (6, 1007) CYCLE, PREMIN, IMIN
179*             IF (PREMIN .LE. 0.0D0)  STOP
180*             CALL MAXMIN (PRE(2), N, PREMAX, IMAX, PREMIN, IMIN)
181*             IF (PREMIN .LE. 0.0D0) WRITE (6, 1008) CYCLE, PREMIN, IMIN
182*             IF (PREMIN .LE. 0.0D0)  STOP
183* C
184* C       ESTABLISH INTEGRATION AND CONTROL CONSTANTS FOR THIS CYCLE.
185*             N1 = N + 1
186*             CALL MAXMIN (MASSC(2), N, AMXMAS, IMAX, AMNMAS, IMIN)
187*             ERROR = DSQRT(AMXMAS/AMNMAS)*ERRLIM
188*             RROLD = RAD(N1)
189*             RLOLD = RAD(1)
190*             VROLD = VEL(N1)
191*             VLOLD = VEL(1)
192*             NCALL = NCALL + 1
193*             NCALIM = MINO (NCALL, NDAMP)
194*             EPSR = DFLOAT(NCALIM)*EPSRO/DFLOAT(NDAMP)
195*             EPSV = DFLOAT(NCALIM)*EPSVO/DFLOAT(NDAMP)
196*             OMEPSR = 1.0D0 - EPSR
197*             OMEPSV = 1.0D0 - EPSV
198* C
199* C       CHECK THE TIMESTEP FOR SUBCYCLING AND NOTE ANY PROBLEMS.
200*             CALL DTFLOW (RAD, VEL, DTVAL, N)
201*             NSTEP = 1
202*             DT = DTIN
203*             FGOHED = 0.999D0*FGOHED
204*             IF (DTVAL .GT. DTIN) GO TO 40
205*             IF (DTVAL .LT. 0.01D0*DTIN) GO TO 30
206*             WRITE (6, 1002) CYCLE, DTVAL, DTIN
207*             NSTEP = (DTIN + DTVAL)/DTVAL
208*             DT = DTIN/FLOAT(NSTEP)
209*             GO TO 40
210*      30      WRITE (6, 1003) CYCLE, DTVAL, DTIN
```

```
211*          STOP
212* C
213* C    INITIALIZE VARIABLES FOR THE SUBCYCLING AND ITERATIONS.
214*    40    DVROLD = (VRNEW - VROLD)/FLOAT(NSTEP)
215*          DVLOLD = (VLNEW - VLOLD)/FLOAT(NSTEP)
216*          DRROLD = (RRNEW - RROLD)/FLOAT(NSTEP)
217*          DRLOLD = (RLNEW - RLOLD)/FLOAT(NSTEP)
218*          DO 50 I = 2, N
219*          RIN(I) = RAD(I) + DT*VEL(I)
220*    50    VIN(I) = VEL(I)
221*          RIN(1) = RLOLD + DRLOLD
222*          RIN(N1) = RROLD + DRROLD
223*          VIN(1) = VLOLD + DVLOLD
224*          VIN(N1) = VROLD + DVROLD
225*          DO 55 I = 2, N1
226*          PREO(I) = PRE(I)
227*    55    PREN(I) = PRE(I)
228*          CALL USEGEO (RAD, AREAO, RADCO, LAMO, N)
229*          CALL USEEOS (RHOO, PREO, LAMEOS, DLAMDP, N)
230*          CALL USEGEO (RIN, AREAN, RADCN, LAMN, N)
231*          CALL USEEOS (RHON, PREN, LAMEOS, DLAMDP, N)
232*          RPREMX = 1.0D0/PREMAX
233*          DO 60 I = 2, N1
234*    60    PRE(I) = 0.0D0
235* C
236* C    PERFORM TIMESTEP SUBCYCLING
237*          DO 550 ISTEP = 1, NSTEP
238*          NTIME = NTIME + 1
239* C
240* C    PERFORM THE ITERATION FOR NEW VALUES RHON, LAMN, PREN, RIN AND VIN.
241*          DO 500 ITER = 1, ITEMAX
242*          NITER = NITER + 1
243* C
244* C    CALCULATE QUANTITIES USED IN TRIDIAGONAL EXPRESSIONS.
245*          DO 200 I = 1, N1
246*          AREAH(I) = 0.5D0*(AREAO(I) + AREAN(I))
247*   200    RBARI(I) = 0.5D0*(RAD(I) + RIN(I))
248*          DO 205 I = 2, N1
249*          DLAM(I) = LAMEOS(I) - LAMN(I)
250*   205    RBAR(I) = 0.5D0*(RADCO(I) + RADCN(I))
251*          DO 210 I = 2, N
252*          RDRI(I) = RHON(I)*(RBARI(I) - RBAR(I)) + RHON(I+1)*(RBAR(I+1)
253*     1             - RBARI(I))
254*   210    DTORHO(I) = DT/RDRI(I)
255* C
256* C    CALCULATE EXPRESSIONS USED IN THE TRIDIAGONAL COEFFICIENTS.
257*          OMERDT = -OMEPSR*DT
258*          DO 250 I = 2, N
259*          AI(I) = VEL(I) - DTORHO(I)*EPSV*(PREO(I+1) - PREO(I))
260*   250    BI(I) = DTORHO(I)*OMEPSV
261*          DO 260 I = 2, N1
262*          DP(I) = OMERDT*AREAH(I)
263*          DM(I) = OMERDT*AREAH(I-1)
```

```
264*            C(I) = PREN(I)*DLAMDP(I) + DP(I)*VIN(I) - DM(I)*VIN(I-1)
265*      260   C(I) = C(I) -DLAM(I)
266* C
267* C       CALCULATE THE TRIDIAGONAL COEFFICIENTS WITH BOUNDARY CONDITIONS.
268*            DO 300 I = 2, N1
269*            DD(I) = C(I)
270*      300   BB(I) = DLAMDP(I)
271*            DO 305 I = 2, N
272*            CC(I) = -DP(I)*BI(I)
273*            AA(I+1) = -DM(I+1)*BI(I)
274*            BB(I) = BB(I) - CC(I)
275*      305   DD(I) = DD(I) - DP(I)*AI(I)
276*            DO 310 I = 3, N1
277*            BB(I) = BB(I) - AA(I)
278*      310   DD(I) = DD(I) + DM(I)*AI(I-1)
279*            DD(2) = DD(2) + DM(2)*VLNEW
280*            DD(N1) = DD(N1) - DP(N1)*VRNEW
281*            AA(2) = 0.0D0
282*            CC(N1) = 0.0D0
283* C
284* C       SOLVE THE TRIDIAGONAL SYSTEM AND CONSTRUCT THE NEW VALUES AT TIME
285* C       T + DT FOR THE CURRENT ITERATION. TRIDDV FAILS WHEN N < 15 THUS
286* C       TRIDDS, THE SCALAR VERSION, IS USED IN THIS REGIME INSTEAD.
287*            IF (N .LT. 15) CALL TRIDDS (N, AA(2), BB(2), CC(2), DD(2),
288*      1         PREN(2), SCA(2), SCB(2))
289*            IF (N .GE. 15) CALL TRIDDV (N, AA(2), BB(2), CC(2), DD(2),
290*      1         PREN(2), SCA(2), SCB(2))
291* C
292* C       IN SOME CIRCUMSTANCES IT MAY BE APPROPRIATE TO LET THE PRESSURE GO
293* C       NEGATIVE. IS SUCH CASES THE FOLLOWING LOOP MUST BE REMOVED OR
294* C       PREMIN MUST BE ALLOWED TO GO NEGATIVE.
295*            DO 340 I = 2, N1
296*      340   PREN(I) = DMAX1 (0.01D0*PREMIN, PREN(I))
297*            DO 350 I = 2, N
298*            VIN(I) = AI(I) - BI(I)*(PREN(I+1) - PREN(I))
299*      350   RIN(I) = RAD(I) + DT*(EPSR*VEL(I) + OMEPSR*VIN(I))
300*            CALL USEGEO (RIN, AREAN, RADCN, LAMN, N)
301*            CALL USEEOS (RHON, PREN, LAMEOS, DLAMDP, N)
302* C
303* C       CHECK ON WHETHER THE ITERATION HAS CONVERGED.
304*            DO 400 I = 2, N1
305*            SCA(I) = PREN(I) - PRE(I)
306*            SCA(I) = DABS(SCA(I))*RPREMX
307*            SCB(I) = ERROR
308*      400   PRE(I) = PREN(I)
309*            NOTCON = 0
310*            ICELL = 0
311*            ERRMAX = 0.0D0
312*            DO 405 I = 2, N1
313*            IF (SCA(I) .LT. SCB(I)) GO TO 405
314*            NOTCON = NOTCON + 1
315*            IF (SCA(I) .GT. ERRMAX) ICELL = I
316*            ERRMAX = DMAX1 (ERRMAX, SCA(I))
```

```
317*    405     CONTINUE
318*            IF (NOTCON .EQ. 0) GO TO 505
319* C
320*    500     CONTINUE
321* C
322* C      PRINT OUT IF WE HAVE NOT CONVERGED.
323*            IF (MOD(CYCLE, 10) .EQ. 0)
324*      1     WRITE (6, 1001) CYCLE, DT, ITEMAX, ERROR, ICELL, ERRMAX,
325*      2          NOTCON
326*    505     CONTINUE
327* C
328* C      SET UP FOR ANOTHER SUBCYCLE TIMESTEP.
329*            IF (ISTEP .EQ. NSTEP) GO TO 550
330*            DO 520 I = 1, N1
331*            AREAO(I) = AREAN(I)
332*            VEL(I) = VIN(I)
333*    520     RAD(I) = RIN(I)
334*            DO 525 I = 1, N1
335*            RIN(I) = RAD(I) + DT*VEL(I)
336*    525     VIN(I) = VEL(I)
337*            RIN(1) = RAD(1) + DRLOLD
338*            RIN(N1) = RAD(N1) + DRROLD
339*            VIN(1) = VEL(1) + DVLOLD
340*            VIN(N1) = VEL(N1) + DVROLD
341*            DO 530 I = 2, N1
342*            RHO(I) = RHON(I)
343*            PREO(I) = PRE(I)
344*    530     RHOO(I) = RHON(I)
345*            CALL USEGEO (RIN, AREAN, RADCN, LAMN, N)
346*            CALL USEEOS (RHON, PREN, LAMEOS, DLAMDP, N)
347*    550     CONTINUE
348* C
349* C      CLEAN UP FOR EXIT.
350*            DO 600 I = 1, N1
351*            VEL(I) = VIN(I)
352*    600     RAD(I) = RIN(I)
353*            DO 601 I = 2, N1
354*    601     RHO(I) = RHON(I)
355*            DO 602 I = 3, N
356*            RHO(I) = (1.0D0-FGOHED)*RHON(I) + FGOHED*(RHON(I+1)+RHON(I-1))
357*    602     PRE(I) = (1.0D0-FGOHED)*PREO(I) + FGOHED*(PREO(I+1)+PREO(I-1))
358*            RETURN
359* C
360* C
361*            ENTRY GOAHED
362*            FGOHED = 0.0
363*            RETURN
364* C
365* C
366*            ENTRY ADINCO (MODE, CYCLE)
367* C      ------------
368* C
369* C      ADINCO PRINTS OUT THE NUMBER OF CALLS TO ADINC, THE NUMBER OF
```

```
370* C     TIMESTEPS INCLUDING SUBCYCLING PERFORMED BY ADINC DURING THOSE
371* C     CALLS, AND THE TOTAL NUMBER OF ITERATIONS PERFORMED SINCE THE LAST
372* C     CALL TO ADINCO.
373* C
374* C     MODE = 0      PRINT OUT AND RESET THE COUNTERS IN ADINC.
375* C     MODE .NE. 0   RESET ALL BUT NCALL COUNTER FOR FILTERING OPERATION
376* C     CYCLE         = TIMESTEP NUMBER FOR IDENTIFICATION PURPOSES
377* C
378*             WRITE (6, 1009) CYCLE, NCALL, NTIME, NITER
379*             IF (MODE .EQ. 0) NCALL = 0
380*             NTIME = 0
381*             NITER = 0
382*             RETURN
383* C
384* C
385*       ENTRY SETEPS (ERO, EVO, MDAMP, EROUT, EVOUT)
386* C     ------------
387* C
388* C     SETEPS PERMITS THE USER TO RESET THE EXPLICITNESS PARAMETERS FOR
389* C     DIFFERENT TYPES OR STAGES OF FLUID PROBLEMS. THE EXPLICITNESS
390* C     PARAMETERS ARE MAPPED INTO THE RANGE 0 <= EPS <= 1.
391* C
392* C     ERO           = THE NEW POSITION EXPLICITNESS PARAMETER
393* C     EVO           = THE NEW VELOCITY EXPLICITNESS PARAMETER
394* C     MDAMP         = THE NEW VALUE OF NDAMP, # OF DAMPING CYCLES
395* C     EROUT         = THE MOST RECENT POSITION EXPLICITNESS PARAMETER
396* C     EVOUT         = THE MOST RECENT VELOCITY EXPLICITNESS PARAMETER
397* C
398*             REAL*8   ERO, EVO, EROUT, EVOUT
399* C
400*             NDAMP = MAXO (MDAMP, 1)
401*             EPSRO = DMIN1 (ERO, 1.0D0)
402*             EPSVO = DMIN1 (EVO, 1.0D0)
403*             EPSRO = DMAX1 (EPSRO, 0.0D0)
404*             EPSVO = DMAX1 (EPSVO, 0.0D0)
405*             EROUT = EPSR
406*             EVOUT = EPSV
407*             RETURN
408*       END
409*       SUBROUTINE SETGEO (ALPHA, GEOMCO)
410* C
411* C
412* C     SETGEO CONTROLS THE GEOMETRIC ASPECTS OF AN ADINC INTEGRATION.
413* C     SETGEO IS CALLED BY THE USER TO TELL THE ADINC PACKAGE EXACTLY
414* C     WHICH OF THE FOUR POSSIBLE GEOMETRIES HE WISHES TO USE.
415* C
416* C     ALPHA = 1      CARTESIAN COORDINATES.
417* C     ALPHA = 2      CYLINDRICAL COORDINATES.
418* C     ALPHA = 3      SPHERICAL COORDINATES.
419* C     ALPHA = 4      POWER SERIES COORDINATES.
420* C     GEOMCO(1..5) = ARRAY OF FIVE COEFFICIENTS IN THE EXPRESSION FOR
421* C                    CELL AREA AND INTEGRATED VOLUME USED BELOW.
422* C
```

```
423* C    SETGEO INITIALIZES THE QUANTITIES GALPHA, HALPHA, G(I), AND H(I)
424* C    FOR LATER REPEATED USE IN ENTRY USEGEO BELOW.
425* C
426*          PARAMETER NPT = 202
427*          INTEGER  ALPHA,      ALPHH
428*          REAL*8   GALPHA,     HALPHA,     G(5),       H(5)
429*          REAL*8   GEOMCO(5),  RIAM1(NPT), DELR(NPT),  TVOL(NPT)
430*          REAL*8   DTMIN(NPT), ABSV(NPT),  DRMIN,      PI
431*          REAL*8   DAMAX
432*          EQUIVALENCE (DTMIN(1), TVOL(1)),  (ABSV(1), DELR(1))
433*          EQUIVALENCE (RIAM1(1), TVOL(1))
434* C
435* C    CHECK THE INPUT TO SETGEO AND INITIALIZE.
436*          IF (ALPHA.LT.1 .OR. ALPHA.GT.4) WRITE (6, 1001) ALPHA, GEOMCO
437*          IF (ALPHA.LT.1 .OR. ALPHA.GT.4) STOP
438* 1001     FORMAT ('0SETGEO INPUT PROBLEM. ALPHA OUT OF RANGE.
439*     1       I4, 2X, 1P5D12.4)
440*          PI = 3.14159265358979D0
441*          ALPHH = ALPHA
442*          GO TO (10, 20, 30, 40), ALPHH
443* C
444* C    MODE = 1     RESET ADINC FOR CARTESIAN COORDINATES.
445*    10    GALPHA = 1.0D0
446*          HALPHA = 1.0D0
447*          RETURN
448* C
449* C    MODE = 2     RESET ADINC FOR CYLINDRICAL COORDINATES.
450*    20    GALPHA = PI
451*          HALPHA = 2.0D0*PI
452*          RETURN
453* C
454* C    MODE = 3     RESET ADINC FOR SPHERICAL COORDINATES.
455*    30    GALPHA = 4.0D0*PI/3.0D0
456*          HALPHA = 4.0D0*PI
457*          RETURN
458* C
459* C    MODE = 4     RESET ADINC FOR POWER SERIES COORDINATES.
460*    40    DO 41 I = 1, 5
461*          G(I) = GEOMCO(I)
462*    41    H(I) = G(I)/DFLOAT(I)
463*          RETURN
464* C
465* C
466*          ENTRY USEGEO (RAD, AREA, RADC, LAMC, N)
467* C        ------------
468* C
469* C    GIVEN A MONOTONICALLY INCREASING SET OF CELL INTERFACE POSITIONS,
470* C    THE INTERFACE AREAS, CELL CENTER LOCATIONS, AND CELL VOLUMES ARE
471* C    CALCULATED IN A FULLY VECTORIZED MANNER. THIS GEOMETRIC UTILITY
472* C    IS USED BY ADINC, DIAGNOSTICS ROUTINES AND THE MAIN PROGRAM -
473* C    WHENEVER THE CELL INTERFACE CONFIGURATION IS CHANGED - TO UPDATE
474* C    THE GEOMETRIC QUANTITIES.
475* C
```

```
476* C      RAD(I)        = POSITION OF THE I-TH INTERFACE (I = 1, N+1) (CM)
477* C      AREA(I)       = AREA IN THE COMPUTATIONAL DOMAIN OF THE I-TH CELL
478* C                      INTERFACE (CM**2)
479* C      RADC(I)       = POSITION OF THE I-TH CELL CENTER (I = 2, N+1) (CM)
480* C      LAMC(I)       = VOLUME OF CELL I BETWEEN INTERFACES I, I+1. (CM**3)
481* C      N             = NUMBER OF INTERIOR CELLS IN THE SYSTEM
482* C
483*          REAL*8   RAD(NPT),   AREA(NPT),   RADC(NPT),   LAMC(NPT)
484* C
485* C
486* C      CHECK THE INPUT TO USEGEO FOR REASONABLENESS.
487*          NPTO = NPT
488*          IF (N.LE.1 .OR. N.GT.NPT-2) WRITE (6, 1002) N, NPTO
489*          IF (N.LE.1 .OR. N.GT.NPT-2) STOP
490*          DO 50 I = 1, N
491*    50    TVOL(I+1) = RAD(I+1) - RAD(I)
492*          CALL MAXMIN (TVOL(2), N, TVOL(1), IMAX, DRMIN, IMIN)
493*          IF (DRMIN .LE. 0.0D0) WRITE (6, 1003) DRMIN, IMIN
494*          IF (DRMIN .LE. 0.0D0) STOP
495*  1002    FORMAT ('0USEGEO INPUT PROBLEM. N OUT OF RANGE. ', I4, I4,
496*        1          ' CALCULATION STOPPED.')
497*  1003    FORMAT ('0USEGEO INPUT PROBLEM. CELL SIZE NEGATIVE ',
498*        1          1PD12.4, ' AT CELL ', I4, '  CALCULATION STOPPED.')
499* C
500*          NP = N + 1
501*          GO TO (100, 200, 300, 400), ALPHH
502* C
503*   100    DO 101 I = 1, NP
504*   101    RIAM1(I) = 1.0D0
505*          GO TO 500
506* C
507*   200    DO 201 I = 1, NP
508*   201    RIAM1(I) = RAD(I)
509*          GO TO 500
510* C
511*   300    DO 301 I = 1, NP
512*   301    RIAM1(I) = RAD(I)*RAD(I)
513* C
514* C      FOR THE REGULAR GEOMETRIES CALCULATE THE AREA AND VOLUME.
515*   500    DO 501 I = 1, NP
516*          AREA(I) = HALPHA*RIAM1(I)
517*   501    TVOL(I) = GALPHA*RIAM1(I)*RAD(I)
518*          GO TO 600
519* C
520* C      FOR THE POWER SERIES (NOZZLE) COORDINATES THE USER MUST SPECIFY
521* C      ALL OF THE GEOMETRIC COEFFICIENTS VIA THE INITIALIZING ARRAY
522* C      GEOMCO IN THE CALL TO SETGEO. HERE G(1) = GEOMCO(1), ETC.
523* C      AREA(R) = G1 + G2*R + G3*R**2 + G4*R**3 + G5*R**4
524* C      TVOL(R) = H1*R + H2*R**2 + H3*R**3 + H4*R**4 + H5*R**5
525*   400    DO 401 I = 1, NP
526*          AREA(I) = G(5)*RAD(I)
527*          TVOL(I) = H(5)*RAD(I)
528*          AREA(I) = RAD(I)*(AREA(I) + G(4))
```

```
529*             TVOL(I) = RAD(I)*(TVOL(I) + H(4))
530*             AREA(I) = RAD(I)*(AREA(I) + G(3))
531*             TVOL(I) = RAD(I)*(TVOL(I) + H(3))
532*             AREA(I) = RAD(I)*(AREA(I) + G(2))
533*             TVOL(I) = RAD(I)*(TVOL(I) + H(2))
534*             AREA(I) = AREA(I) + G(1)
535*     401     TVOL(I) = RAD(I)*(TVOL(I) + H(1))
536* C
537* C       COMPUTE THE CELL VOLUME AND CELL CENTER LOCATIONS.
538*     600     CALL MAXMIN (AREA(1), NP, DAMAX, IMAX, DRMIN, IMIN)
539*             IF (DRMIN .LE. 0.0D0) WRITE (6, 2001) DRMIN, IMIN
540*             IF (DRMIN .LE. 0.0D0) STOP
541*    2001     FORMAT ('0USEGEO PROBLEM, NEGATIVE AREA ', 1PD12.4,
542*       1        ' AT CELL ', I4, ' CALCULATION STOPPED.')
543*             DO 601 I = 2, NP
544*             LAMC(I) = TVOL(I) - TVOL(I-1)
545*     601     RADC(I) = (RAD(I)*AREA(I-1) + RAD(I-1)*AREA(I))/
546*       1        (AREA(I) + AREA(I-1))
547*             RETURN
548* C
549* C
550*             ENTRY DTFLOW (ROD, VEL, DTVAL, N)
551* C           ------------
552* C
553* C       DTFLOW CALCULATES A PERMISSIBLE TIMESTEP DTVAL GIVEN THE SET OF
554* C       N + 1 CELL INTERFACES AND THEIR VELOCITIES.
555* C
556* C       ROD(I)      = POSITION OF THE I-TH CELL INTERFACE (CM)
557* C       VEL(I)      = VELOCITY OF THE I-TH CELL INTERFACE (CM/SEC)
558* C       DTVAL       = THE ESTIMATED VALUE OF A PERMISSIBLE TIMESTEP WHICH
559* C                     PREVENTS INTERFACE CROSSING ASSUMING THE MOTION GIVEN
560* C       N           = THE NUMBER OF INTERIOR CELLS IN THE CURRENT SYSTEM
561* C
562*             REAL*8   ROD(N), VEL(N), DTVAL, EPS, DVSAFE, DTMAX
563* C
564* C
565* C       CHECK THE INPUT TO DTFLOW FOR REASONABLENESS.
566*             NPTO = NPT
567*             IF (N.LE.2 .OR. N.GT.NPT-2) WRITE (6, 1004) N, NPTO
568*             IF (N.LE.2 .OR. N.GT.NPT-2) STOP
569*             N1 = N + 1
570*             EPS = 0.49D0
571*             DO 715 I = 2, N1
572*     715     DELR(I) = ROD(I) - ROD(I-1)
573*             CALL MAXMIN (DELR(2), N, DELR(1), IMAX, DRMIN, IMIN)
574*             IF (DRMIN .LE. 0.0D0) WRITE (6, 1005) DRMIN, IMIN
575*             IF (DRMIN .LE. 0.0D0) STOP
576*    1004     FORMAT ('0DTFLOW INPUT PROBLEM. N OUT OF RANGE. ',  I4, I4,
577*       1        ' CALCULATION STOPPED.')
578*    1005     FORMAT ('0DTFLOW INPUT PROBLEM. CELL SIZE NEGATIVE ',
579*       1        1PD12.4, ' AT CELL ', I4, ' CALCULATION STOPPED.')
580* C
581* C       REQUIRE VEL*DT < MINIMUM OF CELL WIDTHS DELR.
```

```
582*            DO 720 I = 2, N
583*      720   DTMIN(I) = EPS*DMIN1(DELR(I), DELR(I+1))
584*            DTMIN(1) = DELR(2)
585*            DTMIN(N1) = DELR(N1)
586*            DO 725 I = 1, N1
587*      725   ABSV(I) = DABS(VEL(I))
588*            DVSAFE = 1.0D-40
589*            DO 730 I = 1, N1
590*      730   DTMIN(I) = DTMIN(I)/(ABSV(I) + DVSAFE)
591*            CALL MAXMIN (DTMIN, N1, DTMAX, IMAX, DTVAL, IMIN)
592*            RETURN
593*            END
594*            SUBROUTINE SETMAT (MATER, MASS, GAMMA, RHOCON, N)
595* C
596* C              THIS ROUTINE PROVIDES A VECTORIZED DOUBLE PRECISION EQUATION
597* C      OF STATE CALCULATION FOR ADINC. WHEN THE EQUATION OF STATE IS
598* C      CHANGED, A NUMBER OF OTHER ROUTINES MUST BE MODIFIED AS WELL.
599* C
600* C      EQUATION OF STATE ...     RHO = RHOC + (PRE/ENTC)**(1/GAMMAC).
601* C      MATER(I)    = CELL IDENTIFIER = L.MM WHERE 0 < L < 10 IS THE
602* C                                      LAYER NUMBER AND 0 < MM < 100 IS
603* C                                            THE MATERIAL IDENTIFIER.
604* C      MASSC(I)    = CELL MASS = RHO(I)*LAM(I)
605* C      GAMMAC(I)   = CELL ADIABATIC GAS CONSTANT - FIXED DURING ADINC
606* C      ENTC(I)     = CELL ENTROPY - CONSTANT DURING ADINC HYDRODYNAMICS
607* C      RHOC(I)     = DENSITY CONSTANT IN THE EQUATION OF STATE
608* C
609*            PARAMETER   NPT = 202
610*            REAL*8  DELR(NPT),  DELV(NPT),  DRMAX,          DRMIN
611*            REAL*8  MATER(NPT), MASS(NPT),  GAMMA(NPT), RHOCON(NPT)
612*            REAL*8  MASSC(NPT), ENTC(NPT),  RHOC(NPT),  MATERC(NPT)
613*            REAL*8  GAMMAC(NPT),            DELRHO(NPT)
614*            COMMON   /ADICOM/  MATERC, MASSC, GAMMAC, ENTC, RHOC, NCELLS
615* C
616* C
617* C      CHECK THE INPUT TO SETMAT FOR REASONABLENESS.
618*            NPTO = NPT
619*            IF (N.LE.2 .OR. N.GT.NPT-2) WRITE (6, 1001) N, NPTO
620*            IF (N.LE.2 .OR. N.GT.NPT-2) STOP
621*            CALL MAXMIN (MATER(2), N, DRMAX, IMAX, DRMIN, IMIN)
622*            IF (DRMIN .LE. 0.0D0) WRITE (6, 1002) DRMIN, IMIN
623*            IF (DRMIN .LE. 0.0D0) STOP
624*            CALL MAXMIN (MASS(2), N, DRMAX, IMAX, DRMIN, IMIN)
625*            IF (DRMIN .LE. 0.0D0) WRITE (6, 1003) DRMIN, IMIN
626*            IF (DRMIN .LE. 0.0D0) STOP
627*            CALL MAXMIN (RHOCON(2), N, DRMAX, IMAX, DRMIN, IMIN)
628*            IF (DRMIN .LT. 0.0D0) WRITE (6, 1004) DRMIN, IMIN
629*            IF (DRMIN .LT. 0.0D0) STOP
630* 1001     FORMAT ('0SETMAT INPUT PROBLEM. N OUT OF RANGE. ', I4, I4,
631*      1           ' CALCULATION STOPPED.')
632* 1002     FORMAT ('0SETMAT INPUT PROBLEM. CELL MATERIAL NEGATIVE ',
633*      1           1PD12.4, ' AT CELL ', I4, ' CALCULATION STOPPED.')
634* 1003     FORMAT ('0SETMAT INPUT PROBLEM. CELL MASS NEGATIVE ',
```

```
635*      1         1PD12.4, ' AT CELL ', I4, '  CALCULATION STOPPED.')
636*   1004      FORMAT ('OSETMAT INPUT PROBLEM. CELL RHOCON NEGATIVE ',
637*      1         1PD12.4, ' AT CELL ', I4, '  CALCULATION STOPPED.')
638* C
639*             NP = N + 1
640*             DO 100 I = 2, NP
641*             MATERC(I) = MATER(I)
642*             MASSC(I) = MASS(I)
643*             GAMMAC(I) = GAMMA(I)
644*   100       RHOC(I) = RHOCON(I)
645*             NCELLS = N
646*             RETURN
647* C
648* C
649*             ENTRY SETEOS (RHO, PRE, N)
650* C           ------------
651* C
652* C     SETEOS COMPUTES THE ENTROPY CELL CONSTANTS (ENTC) GIVEN KNOWN
653* C     VALUES OF THE DENSITY AND PRESSURE IN THE CELLS. SETEOS IS USED
654* C     AT THE BEGINNING OF CALCULATIONS FOR INITIALIZATION AND DURING
655* C     CALCULATIONS WHENEVER NON-IDEAL, SOURCE, OR DISSIPATIVE EFFECTS
656* C     HAVE CHANGED THE CELL ENTROPIES. SINCE ADINC IS PREDICATED ON THE
657* C     CONSTANCY OF THE LAGRANGIAN CELL ENTROPY DURING A FLUID TIMESTEP,
658* C     THIS AMOUNTS TO A FORM OF TIMESTEP SPLITTING.
659* C
660* C     RHO(I)      = DENSITY OF MATERIAL IN CELL I (I = 2, N+1) (GM/CC)
661* C                 THESE QUANTITIES ARE GIVEN AS INPUTS TO SETEOS
662* C     PRE(I)      = PRESSURE GIVEN IN CELL I BETWEEN INTERFACES I AND
663* C                 I-1.  THESE VALUES ARE INPUT TO SETEOS (ERG/CC)
664* C     N           = NUMBER OF INTERIOR CELLS IN THE SYSTEM
665* C
666*             REAL*8   RHO(NPT),   PRE(NPT)
667* C
668* C
669* C     CHECK THE INPUT TO SETEOS FOR REASONABLENESS.
670*             NPTO = NPT
671*             IF (N.LE.2 .OR. N.GT.NPT-2) WRITE (6, 1005) N, NPTO
672*             IF (N.LE.2 .OR. N.GT.NPT-2) STOP
673*             CALL MAXMIN (RHO(2), N, DRMAX, IMAX, DRMIN, IMIN)
674*             IF (DRMIN .LE. 0.0D0) WRITE (6, 1006) DRMIN, IMIN
675*             IF (DRMIN .LE. 0.0D0) STOP
676*             CALL MAXMIN (PRE(2), N, DRMAX, IMAX, DRMIN, IMIN)
677*             IF (DRMIN .LE. 0.0D0) WRITE (6, 1007) DRMIN, IMIN
678*             IF (DRMIN .LE. 0.0D0) STOP
679*   1005      FORMAT ('OSETEOS INPUT PROBLEM. N OUT OF RANGE. ',  I4, I4,
680*      1         ' CALCULATION STOPPED.')
681*   1006      FORMAT ('OSETEOS INPUT PROBLEM. CELL DENSITY NEGATIVE ',
682*      1         1PD12.4, ' AT CELL ', I4, '  CALCULATION STOPPED.')
683*   1007      FORMAT ('OSETEOS INPUT PROBLEM. CELL PRESSURE NEGATIVE ',
684*      1         1PD12.4, ' AT CELL ', I4, '  CALCULATION STOPPED.')
685* C
686*             NP = N + 1
687*             DO 200 I = 2, NP
```

```
688*   200      DELRHO(I) = DMAX1 (1.0D-30, (RHO(I) - RHOC(I)))
689*            CALL DUBLOG (DELRHO(2), ENTC(2), N)
690*            DO 210 I = 2, NP
691*   210      ENTC(I) = ENTC(I)*GAMMAC(I)
692*            DO 215 I = 2, NP
693*   215      ENTC(I) = DEXP(ENTC(I))
694*            DO 220 I = 2, NP
695*   220      ENTC(I) = PRE(I)/ENTC(I)
696*            RETURN
697* C
698* C
699*            ENTRY USEEOS (RHO, PRE, LAMEOS, DLAMDP, N)
700* C          ------------
701* C
702* C      USEEOS TAKES THE SET OF CELL PRESSURES (PRE)  AND CALCULATES THE
703* C      CELL DENSITIES EXPECTED FOR THAT PRESSURE BASED ON THE EQUATION
704* C      OF STATE CONSTANTS FOR THE CELL STORED IN COMMON BLOCK /ADICOM/.
705* C      THE EXPECTED CELL VOLUME (LAMEOS) IS COMPUTED FOR EACH CELL AS IS
706* C      THE JACOBEAN DERIVATIVE DLAMDP.
707* C
708* C      RHO(I)        = DENSITY OF MATERIAL IN CELL I (I = 2, N+1) (GM/CC)
709* C                      THESE QUANTITIES ARE COMPUTED AS OUTPUTS OF USEEOS
710* C      PRE(I)        = PRESSURE GIVEN IN CELL I BETWEEN INTERFACES I AND
711* C                      I-1.   THESE VALUES ARE INPUT TO USEEOS (ERG/CC)
712* C      LAMEOS(I)     = THE CELL VOLUME REQUIRED BY THE EQUATION OF STATE
713* C                      GIVEN THE CELL PRESSURE PRE(I) AND E.O.S. CONSTANTS
714* C      DLAMDP(I)     = RATE OF CHANGE OF CELL VOLUME WITH PRESSURE IN THE
715* C                      EQUATION OF STATE
716* C      N             = NUMBER OF INTERIOR CELLS IN THE SYSTEM
717* C
718*            REAL*8   LAMEOS(NPT),            DLAMDP(NPT)
719* C
720* C
721* C      CHECK THE INPUT TO USEEOS FOR REASONABLENESS.
722*            NPTO = NPT
723*            IF (N.LE.2 .OR. N.GT.NPT-2) WRITE (6, 1008) N, NPTO
724*            IF (N.LE.2 .OR. N.GT.NPT-2) STOP
725*            CALL MAXMIN (PRE(2), N, DRMAX, IMAX, DRMIN, IMIN)
726*            IF (DRMIN .LE. 0.0D0) WRITE (6, 1009) DRMIN, IMIN
727*            IF (DRMIN .LE. 0.0D0) STOP
728*  1008      FORMAT ('0USEEOS INPUT PROBLEM. N OUT OF RANGE. ', I4, I4,
729*      1            ' CALCULATION STOPPED.')
730*  1009      FORMAT ('0USEEOS INPUT PROBLEM. CELL PRESSURE NEGATIVE ',
731*      1         1PD12.4, ' AT CELL ', I4, '  CALCULATION STOPPED.')
732* C
733*            N1 = N + 1
734*            DO 300 I = 2, N1
735*   300      DELV(I) = DMAX1(PRE(I),0.0D0)/ENTC(I)
736*            CALL DUBLOG (DELV(2), DELR(2), N)
737*            DO 320 I = 2, N1
738*   320      DELR(I) = DELR(I)/GAMMAC(I)
739*            DO 340 I = 2, N1
740*   340      DELR(I) = DEXP(DELR(I))
741* C
742* C      EVALUATE THE EQUATION OF STATE   RHO = RHOC + (PRE/ENTC)**1/GAMMAC
743*            DO 360 I = 2, N1
744*            RHO(I) = RHOC(I) + DELR(I)
745*   360      LAMEOS(I) = MASSC(I)/RHO(I)
746*            DO 380 I = 2, N1
747*   380      DLAMDP(I) = - LAMEOS(I)*DELR(I)/(RHO(I)*PRE(I)*GAMMAC(I))
748*            RETURN
749*        END
```

Appendix D

```
 1*  C
 2*  C
 3*  C     BSM1, BSM2, AND BSM3 ARE ASC FORTRAN SUBROUTINES FOR EXPLICITLY
 4*  C     SOLVING THE GENERAL ELLIPTIC FIVE-POINT DIFFERENCE EQUATION
 5*  C
 6*  C         AX(I,J)*X(I-1,J-1) + AY(I,J)*X(I,J-1) + BB(I,J)*X(I,J)
 7*  C
 8*  C           + CX(I,J)*X(I+1,J) + CY(I,J)*X(I+1,J+1) = F(I,J)
 9*  C
10*  C     ON AN ARBITRARY MESH.  HERE X IS THE SOLUTION BEING SOUGHT AND
11*  C     F IS AN ARBITRARY FORCING FUNCTION.
12*  C
13*  C
14*  C                         UTILIZATION
15*  C
16*  C     BSM1      - COMPUTES THE INFLUENCE MATRIX (INDEPENDENT OF THE
17*  C                 FORCING FUNCTION AND BOUNDARY CONDITIONS).  IF THE
18*  C                 COEFFICIENTS AX, AY, BB, CX, CY ARE FIXED, BSM1 IS
19*  C                 CALLED ONLY ONCE.
20*  C
21*  C     BSM2      - CALCULATES THE SOLUTION X USING THE INFLUENCE MATRIX,
22*  C                 FORCING FUNCTION F, AND BOUNDARY CONDITIONS, BY THE SEVP
23*  C                 METHOD.
24*  C
25*  C     BSM3      - CALCULATES CORRECTIONS.
26*  C
27*  C     AUXILIARY ROUTINES CALLED BY THE ABOVE:
28*  C
29*  C     MATINV   -  GENERAL MATRIX INVERSION ROUTINE
30*  C
31*  C
32*  C                         CALLING SEQUENCE
33*  C
34*  C     AX, AY, BB, CX, CY - COEFFICIENTS OF EQUATION TO BE SOLVED (SEE
35*  C                          EQUATION ABOVE)
36*  C
37*  C     RINV, RINV1        - INFLUENCE MATRICES
38*  C
39*  C     RCOR               - SCRATCH ARRAY
40*  C
41*  C     DUMMY1             - SCRATCH ARRAY (MUST BE EQUIVALENCED WITH LA
42*  C                          N2 X N2 ELEMENTS OF RINV1)
43*  C
44*  C     NBSIZ2             - NUMBER OF INTERIOR POINTS IN THE MARCHING (X)
45*  C                          DIRECTION OF A BLOCK
46*  C
47*  C     IS, IE             - INDICES OF FIRST, LAST ELEMENTS OF BLOCK IN
48*  C                          MARCHING DIRECTION
49*  C
50*  C     F                  - FORCING FUNCTION
51*  C
```

```
52* C    X                  - SOLUTION
53* C
54* C    RTILDA             - SCRATCH ARRAY
55* C
56* C    ERROR              - LIMIT ON MAXIMUM RELATIVE ERROR IN SOLUTION
57* C
58* C    SUMF               - SCRATCH ARRAY
59* C
60* C    F11, F1N, F21, F2M - ARRAYS USED TO PRESCRIBE BOUNDARY CONDITIONS
61* C
62* C    AI1, AIN, A21, A2M - SWITCHES FOR DIRICHLET/NEUMANN BOUNDARY
63* C                         CONDITIONS
64* C
65* C
66* C    IN THE VERSION CURRENTLY IMPLEMENTED, THE DIMENSIONS OF X, F, AND
67* C    THE COEFFICIENT ARRAYS ARE SET INSIDE AND OUTSIDE THE SOLVER USING
68* C    PARAMETER STATEMENTS.  THE DIMENSION IN THE X DIRECTION IS M AND
69* C    THE DIMENSION IN THE Y DIRECTION IS N.  THE NUMBER OF INTERIOR
70* C    POINTS IN THE Y DIRECTION IS N2 = N - 2.  THE NUMBER OF BLOCKS,
71* C    NBLK, IS ALSO DEFINED IN PARAMETER STATEMENTS.  RINV IS DIMENSIONED
72* C    (N2, N2, NBLK); RINV1 IS DIMENSIONED (N2, N2, NBLK1), WHERE
73* C    NBLK1 = NBLK - 1;  AX, AY, BB, CX, CY, X, AND F ARE DIMENSIONED
74* C    (M, N); THE OTHER ARRAYS ARE DIMENSIONED ACCORDING TO RCOR(3, N)
75* C    NBSIZ2(NBLK), IS(NBLK), IE(NBLK), SUMF(NBLK), RTILDA(N2), DUMMY1(N2, N2),
76* C    F11(M), F1N(M), F21(N), F2M(N).
77*
78* C
79* C    THE BOUNDARY CONDITIONS ARE SET PRIOR TO CALLING BSM2, ACCORDING TO
80* C
81* C        X(I,1) = (1 - A11)*X(I,1) + A11*(X(I,2) - F11(I))
82* C    AND
83* C        X(I,N) = (1 - A1N)*X(I,N) + A1N*(X(I,N-1) - F1N(I))
84* C
85* C    FOR 1 <= I <= M, AND
86* C
87* C        X(1,J) = (1 - A21)*X(1,J) + A21*(X(2,J) - F21(J))
88* C    AND
89* C        X(M,J) = (1 - A2M)*X(M,J) + A2M*(X(M-1,J) + F2M(J))
90* C
91* C    FOR 1 <= J <= N, WHERE THE A'S  EQUAL 0 FOR DIRICHLET BOUNDARY
92* C    CONDITIONS AND 1 FOR NEUMANN BOUNDARY CONDITIONS.
93* C
94* C    AFTER DEFINING PARAMETERS, DECLARING ARRAYS, EQUIVALENCING DUMMY1 AND
95* C    RINV1, DEFINING THE COEFFICIENTS AND F, AND SETTING THE BOUNDARY
96* C    CONDITIONS AND ERROR, TWO CALLS ARE NEEDED:
97* C
98* C    CALL BSM1(AX, AY, BB, CX, CY, RINV, RINV1, RCOR, DUMMY1, NBSIZ2, IS, IE)
99* C
100* C   AND
101* C
102* C   CALL BSM2(AX, AY, BB, CX, CY, RINV, RINV1, RCOR, F, X, RTILDA, ERROR,
103* C  1    SUMF, IS, IE, F11, F1N, F21, F2M, AI1, AIN, A21, A2M)
104* C
```

```
105*  C       IS AND IE ARE CALCULATED BY BSM1; X IS RETURNED BY BSM2.
106*  C
107*  C
108*          SUBROUTINE BSM1(AX,AY,BB,CX,CY,RINV,RINV1,RCOR,DUMMY1,
109*         1NBSIZ2,IS,IE)
110*          IMPLICIT REAL*8(A-H,O-Z)
111*          PARAMETER M=51,N=51,M1=M-1,N1=N-1,N2=N-2,NBLK=3,NBLK1=NBLK-1
112*          DIMENSION AX(M,N),AY(M,N),CX(M,N),CY(M,N),BB(M,N)
113*          DIMENSION RINV(N2,N2,NBLK),RINV1(N2,N2,NBLK1),RCOR(3,N)
114*          DIMENSION NBSIZ2(NBLK),IS(NBLK),IE(NBLK)
115*          DIMENSION DUMMY1(N2,N2)
116*          EQUIVALENCE(DUMMY1,RINV(1,1,NBLK))
117*          IE(1)=NBSIZ2(1)+2
118*          DO 90 NB=2,NBLK
119*          IE(NB)=IE(NB-1)+NBSIZ2(NB)+1
120*   90 CONTINUE
121*          DO 95 NB=1,NBLK1
122*          IS(NB+1)=IE(NB)-1
123*   95 CONTINUE
124*          IS(1)=1
125*          DO 100 I=1,N2
126*          DO 100 J=1,N2
127*          DO 100 K=1,NBLK
128*          RINV(I,J,K)=0.0
129*  100 CONTINUE
130*          DO 115 J1=1,N2
131*          DO 110 I=1,3
132*          DO 110 J=1,N
133*          RCOR(I,J)=0.0
134*  110 CONTINUE
135*          RCOR(2,J1+1)=1.0
136*          NBS=IE(1)-1
137*          DO 130 I1=2,NBS
138*          DO 135 J=2,N1
139*          RCOR(3,J)=(-AY(I1,J)*RCOR(2,J-1)-AX(I1,J)*RCOR(1,J)-BB(I1,J)*
140*         1RCOR(2,J)-CY(I1,J)*RCOR(2,J+1))/CX(I1,J)
141*  135 CONTINUE
142*          DO 140 J=1,N
143*          RCOR(1,J)=RCOR(2,J)
144*          RCOR(2,J)=RCOR(3,J)
145*  140 CONTINUE
146*  130 CONTINUE
147*          DO 145 J=1,N2
148*          RINV(J1,J,1)=RCOR(1,J+1)
149*          DUMMY1(J1,J)=RCOR(2,J+1)
150*  145 CONTINUE
151*  115 CONTINUE
152*          CALL MATINV(DUMMY1)
153*          DO 160 I=1,N2
154*          DO 160 J=1,N2
155*          SUM=0.0
156*          DO 161 K=1,N2
157*          SUM=SUM+DUMMY1(I,K)*RINV(K,J,1)
```

```
158*    161 CONTINUE
159*        RINV1(I,J,1)=SUM
160*    160 CONTINUE
161*        DO 170 I=1,N2
162*        DO 170 J=1,N2
163*        RINV(I,J,1)=DUMMY1(I,J)
164*    170 CONTINUE
165*        DO 205 NB=2,NBLK
166*        DO 215 J1=1,N2
167*        DO 210 I=1,3
168*        DO 210 J=1,N
169*        RCOR(I,J)=0.0
170*    210 CONTINUE
171*        DO 220 J=1,N2
172*        RCOR(1,J+1)=RINV1(J1,J,NB-1)
173*    220 CONTINUE
174*        RCOR(2,J1+1)=1.0
175*        IE1=IE(NB-1)
176*        IE2=IE(NB)-1
177*        IF(NB.LT.NBLK) GO TO 232
178*        IE2=IE2-1
179*    232 CONTINUE
180*        DO 230 I1=IE1,IE2
181*        DO 235 J=2,N1
182*        RCOR(3,J)=(-AY(I1,J)*RCOR(2,J-1)-AX(I1,J)*RCOR(1,J)-BB(I1,J)*
183*       1RCOR(2,J)-CY(I1,J)*RCOR(2,J+1))/CX(I1,J)
184*    235 CONTINUE
185*        DO 240 J=1,N
186*        RCOR(1,J)=RCOR(2,J)
187*        RCOR(2,J)=RCOR(3,J)
188*    240 CONTINUE
189*    230 CONTINUE
190*        IF(NB.EQ.NBLK) GO TO 246
191*        DO 245 J=1,N2
192*        RINV(J1,J,NB)=RCOR(1,J+1)
193*    245 CONTINUE
194*        DO 247 J=1,N2
195*        DUMMY1(J1,J)=RCOR(2,J+1)
196*    247 CONTINUE
197*        GO TO 249
198*    246 CONTINUE
199*        DO 248 J=2,N1
200*        DUMMY1(J1,J-1)=AY(M-1,J)*RCOR(2,J-1)+AX(M-1,J)*RCOR(1,J)+BB(M-1,J)
201*       1*RCOR(2,J)+CY(M-1,J)*RCOR(2,J+1)
202*    248 CONTINUE
203*    249 CONTINUE
204*    215 CONTINUE
205*        CALL MATINV(DUMMY1)
206*        IF(NB.EQ.NBLK) GO TO 275
207*        DO 260 I=1,N2
208*        DO 260 J=1,N2
209*        SUM=0.0
210*        DO 261 K=1,N2
```

```
211*          SUM=SUM+DUMMY1(I,K)*RINV(K,J,NB)
212*    261   CONTINUE
213*          RINV1(I,J,NB)=SUM
214*    260   CONTINUE
215*          DO 270 I=1,N2
216*          DO 270 J=1,N2
217*          RINV(I,J,NB)=DUMMY1(I,J)
218*    270   CONTINUE
219*    275   CONTINUE
220*    205   CONTINUE
221*          RETURN
222*          END
223*          SUBROUTINE MATINV(B)
224*          IMPLICIT REAL*8(A-H,O-Z)
225*          PARAMETER M=49
226*          DIMENSION B(M,M)
227*          DIMENSION B1(M),B2(M)
228*          M1=M-1
229*          DO 110 I=1,M1
230*          B1(1)=1.0/B(I,I)
231*          B(I,I)=1.0
232*          DO 112 J=1,M
233*          B(I,J)=B(I,J)*B1(1)
234*    112   CONTINUE
235*          IP1=I+1
236*          DO 120 I1=IP1,M
237*          B1(I1)=B(I1,I)
238*    120   CONTINUE
239*          DO 125 I1=IP1,M
240*          B(I1,I)=0.0
241*    125   CONTINUE
242*          DO 127 J=1,M
243*          B2(J)=B(I,J)
244*    127   CONTINUE
245*          DO 135 I1=IP1,M
246*          DO 135 J=1,M
247*          B(I1,J)=B(I1,J)-B1(I1)*B2(J)
248*    135   CONTINUE
249*    110   CONTINUE
250*          B1(1)=1.0/B(M,M)
251*          B(M,M)=1.0
252*          DO 140 J=1,M
253*          B(M,J)=B(M,J)*B1(1)
254*    140   CONTINUE
255*          DO 150 I=2,M
256*          DO 155 I2=1,I
257*          B1(I2)=B(I2,I)
258*    155   CONTINUE
259*          IM1=I-1
260*          DO 156 I2=1,IM1
261*          B(I2,I)=0.0
262*    156   CONTINUE
263*          DO 157 J=1,M
```

```
264*        B2(J)=B(I,J)
265*  157 CONTINUE
266*        IM1=I-1
267*        DO 160 I2=1,IM1
268*        DO 160 J=1,M
269*        B(I2,J)=B(I2,J)-B1(I2)*B2(J)
270*  160 CONTINUE
271*  150 CONTINUE
272*        RETURN
273*        END
274*        SUBROUTINE BSM2(AX,AY,BB,CX,CY,RINV,RINV1,RCOR,F,X,RTILDA,ERROR,
275*       1SUMF,IS,IE,F11,F1N,F21,F2M,A11,A1N,A21,A2M)
276*        IMPLICIT REAL*8(A-H,O-Z)
277*        PARAMETER M=51,N=51,M1=M-1,N1=N-1,N2=N-2,NBLK=3,NBLK1=NBLK-1
278*        DIMENSION AX(M,N),AY(M,N),CX(M,N),CY(M,N),BB(M,N)
279*        DIMENSION F11(M),F1N(M),F21(N),F2M(N)
280*        DIMENSION X(M,N),F(M,N)
281*        DIMENSION RINV(N2,N2,NBLK),RINV1(N2,N2,NBLK1),RCOR(3,N)
282*        DIMENSION SUMF(NBLK),RTILDA(N2)
283*        DIMENSION IS(NBLK),IE(NBLK)
284*        DO 90 NB=1,NBLK
285*        SUMF(NB)=0.0
286*   90 CONTINUE
287*        DO 95 NB=1,NBLK
288*        DO 95 J=2,N1
289*        SUMF(NB)=SUMF(NB)+DABS(F(IE(NB)-1,J))
290*   95 CONTINUE
291*        DO 96 NB=1,NBLK
292*        IF(SUMF(NB).GT.0.0) GO TO 96
293*        SUMF(NB)=1.0
294*   96 CONTINUE
295*        DO 200 NB=1,NBLK
296*        ISP1=IS(NB)+1
297*        IEM2=IE(NB)-2
298*        DO 205 I=ISP1,IEM2
299*        DO 205 J=2,N1
300*        X(I+1,J)=(F(I,J)-AY(I,J)*X(I,J-1)-AX(I,J)*X(I-1,J)-BB(I,J)*
301*       1X(I,J)-CY(I,J)*X(I,J+1))/CX(I,J)
302*  205 CONTINUE
303*        IF(NB.EQ.NBLK) GO TO 200
304*        DO 522 IT=1,10
305*        I1=IE(NB)-1
306*        DO 215 J=2,N1
307*        RTILDA(J-1)=X(I1+1,J)-(F(I1,J)-AY(I1,J)*X(I1,J-1)-AX(I1,J)*
308*       1X(I1-1,J)-BB(I1,J)*X(I1,J)-CY(I1,J)*X(I1,J+1))/CX(I1,J)
309*  215 CONTINUE
310*        A2=0.0
311*        DO 216 J=1,N2
312*        A2=A2+DABS(RTILDA(J))
313*  216 CONTINUE
314*        A3=A2/SUMF(NB)
315*        IF(A3.LE.1.0E-1) GO TO 230
316*        DO 217 I=1,3
```

```
317*           DO 217 J=1,N
318*           RCOR(I,J)=0.0
319*     217 CONTINUE
320*           DO 223 I=1,N2
321*           DO 223 I1=1,N2
322*           RCOR(2,I+1)=RCOR(2,I+1)+RTILDA(I1)*RINV(I1,I,NB)
323*     223 CONTINUE
324*           IF(NB.EQ.1) GO TO 251
325*           DO 225 I=2,N1
326*           DO 225 K=2,N1
327*           RCOR(1,I)=RCOR(1,I)+RCOR(2,K)*RINV1(K-1,I-1,NB-1)
328*     225 CONTINUE
329*           DO 226 J=2,N1
330*           X(IS(NB),J)=X(IS(NB),J)+RCOR(1,J)
331*     226 CONTINUE
332*     251 CONTINUE
333*           CALL        BSM3(AX,AY,BB,CX,CY,X,RCOR,IS(NB),IE(NB))
334*     522 CONTINUE
335*     230 CONTINUE
336*           DO 220 I=1,3
337*           DO 220 J=1,N
338*           RCOR(I,J)=0.0
339*     220 CONTINUE
340*           DO 224 J=1,N2
341*           DO 224 J1=1,N2
342*           RCOR(2,J+1)=RCOR(2,J+1)+RTILDA(J1)*RINV1(J1,J,NB)
343*     224 CONTINUE
344*           DO 235 J=2,N1
345*           X(IE(NB)-1,J)=X(IE(NB)-1,J)+RCOR(2,J)
346*     235 CONTINUE
347*     501 CONTINUE
348*     200 CONTINUE
349*           DO 300 NB1=1,NBLK
350*           NB=NBLK -NB1+1
351*           ISP1=IS(NB)+1
352*           IEM2=IE(NB)-2
353*           I=IEM2
354*           IF(NB.EQ.NBLK) GO TO 502
355*           DO 305 J=2,N1
356*           X(I+1,J)=(F(I,J)-AY(I,J)*X(I,J-1)-AX(I,J)*X(I-1,J)-BB(I,J)*
357*          1X(I,J)-CY(I,J)*X(I,J+1))/CX(I,J)
358*     305 CONTINUE
359*     502 CONTINUE
360*           DO 552 IT=1,10
361*           IF(NB.EQ.NBLK) GO TO 317
362*           I1=IE(NB)-1
363*           DO 315 J=2,N1
364*           RTILDA(J-1)=X(I1+1,J)-(F(I1,J)-AY(I1,J)*X(I1,J-1)-AX(I1,J)*
365*          1X(I1-1,J)-BB(I1,J)*X(I1,J)-CY(I1,J)*X(I1,J+1))/CX(I1,J)
366*     315 CONTINUE
367*           GO TO 318
368*     317 CONTINUE
369*           DO 319 J=2,N1
```

```
370*          RTILDA(J-1)=F(M-1,J)-(AY(M-1,J)*X(M-1,J-1)+AX(M-1,J)*X(M-2,J)+
371*         1BB(M-1,J)*X(M-1,J)+CY(M-1,J)*X(M-1,J+1))
372*      319 CONTINUE
373*      318 CONTINUE
374*          PRINT 905,(RTILDA(I),I=1,N2)
375*      905 FORMAT(1H ,10E12.5)
376*          A2=0.0
377*          DO 316 J=1,N2
378*          A2=A2+DABS(RTILDA(J   ))
379*      316 CONTINUE
380*          A3=A2/SUMF(NB)
381*          IF(A3.LE.ERROR) GO TO 300
382*          DO 320 I=1,3
383*          DO 320 J=1,N
384*          RCOR(I,J)=0.0
385*      320 CONTINUE
386*          DO 324 J=1,N2
387*          DO 324 J1=1,N2
388*          RCOR(2,J+1)=RCOR(2,J+1)+RTILDA(J1)*RINV(J1,J,NB)
389*      324 CONTINUE
390*          IF(NB.EQ.1) GO TO 551
391*          DO 325 J=2,N1
392*          DO 325 K=2,N1
393*          RCOR(1,J)=RCOR(1,J)           +RCOR(2,K)*RINV1(K-1,J-1,NB-1)
394*      325 CONTINUE
395*          DO 326 J=2,N1
396*          X(IS(NB),J)=X(IS(NB),J)+RCOR(1,J)
397*      326 CONTINUE
398*      551 CONTINUE
399*          CALL          BSM3(AX,AY,BB,CX,CY,X,RCOR,IS(NB),IE(NB))
400*      552 CONTINUE
401*      300 CONTINUE
402*      570 CONTINUE
403*          DO 350 J=2,N1
404*          X(1,J)=(1.0-A21)*X(1,J)+A21*(X(2,J)-F21(J))
405*          X(M,J)=(1.0-A2M)*X(M,J)+A2M*(X(M-1,J)+F2M(J))
406*      350 CONTINUE
407*          DO 360 I=2,M1
408*          X(I,1)=(1.0-A11)*X(I,1)+A11*(X(I,2)-F11(I))
409*          X(I,N)=(1.0-A1N)*X(I,N)+A1N*(X(I,N-1)+F1N(I))
410*      360 CONTINUE
411*          DO 371 J=2,N1
412*          BB(2,J)=BB(2,J)-AX(2,J)*A21
413*          F(2,J)=F(2,J)-AX(2,J)*F21(J)*A21
414*          BB(M-1,J)=BB(M-1,J)-CX(M-1,J)*A2M
415*          F(M-1,J)=F(M-1,J)+CX(M-1,J)*F2M(J)*A2M
416*      371 CONTINUE
417*          DO 372 I=2,M1
418*          BB(I,2)=BB(I,2)-AY(I,2)*A11
419*          F(I,2)=F(I,2)-AY(I,2)*F11(I)*A11
420*          BB(I,N-1)=BB(I,N-1)-CY(I,N-1)*A1N
421*          F(I,N-1)=F(I,N-1)+CY(I,N-1)*F1N(I)*A1N
422*      372 CONTINUE
```

```
423*          RETURN
424*          END
425*          SUBROUTINE BSM3(AX,AY,BB,CX,CY,X,RCOR,IS,IE)
426*          IMPLICIT REAL*8(A-H,O-Z)
427*          PARAMETER M=51,N=51,M1=M-1,N1=N-1,N2=N-2,NBLK=3,NBLK1=NBLK-1
428*          DIMENSION AX(M,N),AY(M,N),BB(M,N),CX(M,N),CY(M,N)
429*          DIMENSION RCOR(3,N),X(M,N)
430*          DO 135 J=2,N1
431*          X(IS+1,J)=X(IS+1,J)+RCOR(2,J)
432*   135 CONTINUE
433*          ISP1=IS+1
434*          IEM2=IE-2
435*          DO 140 I=ISP1,IEM2
436*          DO 145 J=2,N1
437*          RCOR(3,J)=(-AY(I,J)*RCOR(2,J-1)-AX(I,J)*RCOR(1,J)-BB(I,J)*RCOR(2,J
438*         1)-CY(I,J)*RCOR(2,J+1))/CX(I,J)
439*   145 CONTINUE
440*          DO 150 J=2,N1
441*          X(I+1,J)=X(I+1,J)+RCOR(3,J)
442*          RCOR(1,J)=RCOR(2,J)
443*          RCOR(2,J)=RCOR(3,J)
444*   150 CONTINUE
445*   140 CONTINUE
446*          RETURN
447*          END
```

Appendix E

```
 1*  C
 2*  C
 3*  C     ELI AND EXCIT ARE VECTORIZED ASC FORTRAN SUBROUTINES FOR SOLUTION
 4*  C     OF THE NONSELF ADJOINT 5 POINT EQUATION
 5*  C
 6*  C                    (DEL**2 + A*GRAD) P = R
 7*  C
 8*  C     ON A NON-STRETCHED 2D GRID.  HERE DEL**2 IS THE 5 POINT LAPLACIAN
 9*  C     OPERATOR AND A IS A VECTOR WHOSE COMPONENTS ARE SPECIFIED
10*  C     NUMERICALLY ON THE GRID.  P IS THE APPROXIMATE SOLUTION TO BE
11*  C     REFINED ITERATIVELY BY THE CHEBYCHEV METHOD.
12*  C
13*  C
14*  C                         UTILIZATION
15*  C
16*  C     ELI       - REFINE THE SOLUTION ON A COARSE GRID AND THEN A FINE
17*  C                 GRID BY CALLING EXCIT.
18*  C
19*  C     ELI2      - AN ENTRY IN ELI WHICH ALLOWS THE USER TO MAKE MULTIPLE
20*  C                 PASSES THROUGH COARSE AND FINE GRIDS AFTER AN INITIAL
21*  C                 CALL TO ELI.
22*  C
23*  C     EXCIT     - CARRY OUT THE CHEBYCHEV ITERATION AND COMPUTE ERROR.
24*  C
25*  C     AUXILIARY ROUTINES CALLED BY THE ABOVE:
26*  C
27*  C     AV2       - COMPUTE 4 POINT AVERAGES OF FINE GRID QUANTITIES AND
28*  C                 PLACE ON THE COARSE GRID
29*  C
30*  C     IN2       - (AN ENTRY IN AV2) PLACE COARSE GRID QUANTITIES ON THE
31*  C                 FINE GRID THROUGH BILINEAR INTERPOLATION
32*  C
33*  C     PBCSET    - RESET BOUNDARY CONDITIONS ON EXTERIOR GRID POINTS
34*  C                 AFTER EACH ITERATION
35*  C
36*  C     MAX2D     - FIND THE LOCATION OF MAXIMUM ELEMENT OF 2D ARRAY
37*  C
38*  C
39*  C                         CALLING SEQUENCE
40*  C
41*  C     ELI, ELI2, AND EXCIT HAVE THE SAME ARGUMENT LIST:
42*  C
43*  C     CALL ELI (IMAX, NX,DX, NY,DY, P,R, AX,AY, S,ERR)
44*  C
45*  C     IMAX      - NUMBER OF ITERATIONS TO BE PERFORMED ON THE FINE MESH
46*  C
47*  C     NX,NY     - DIMENSIONS OF FINE MESH INCLUDING EXTERIOR GUARD CELLS
48*  C
49*  C     DX, DY    - CONSTANT GRID INTERVALS
50*  C
51*  C     AX, AY    - NOTE11   X & Y COMPONENTS OF VECTOR A DIVIDED BY
```

```
52* C                    2DX AND 2DY RESPECTIVELY.
53* C
54* C     P, R      - APPROXIMATE SOLUTION AND SOURCE TERM.  IF NOTHING
55* C                  BETTER IS AVAILABLE, SET P = 0.
56* C
57* C     S, ERR    - RETURN AS RESIDUAL ERROR AND RMS RESIDUAL/RMS SOURCE.
58* C                  WHEN ELI2 IS CALLED, S MUST HAVE BEEN DEFINED BY A
59* C                  PREVIOUS CALL TO ELI.
60* C
61* C
62* C     THE AUXILIARY COARSE GRID HAS ONLY 1/16 AS MANY POINTS AS THE
63* C     NX,NY GRID AND IS CONTAINED IN COMMON/LMESH/ (DECLARED IN ELI).
64* C
65* C     DEFAULT BOUNDARY CONDITIONS ARE DOUBLY PERIODIC.  TO CHANGE B.C.
66* C     USER SHOULD (1) ALTER PBCSET AND (2) CHANGE EIGENVALUES IN EXCIT
67* C     (STATEMENT 24) AND ELI (CALL TO SIGSET NEAR END).  STEP (2) MAY IN
68* C     SOME CASES BE OMITTED WITHOUT MUCH LOSS IN ACCURACY.
69* C
70* C     IF USER DOES NOT DESIRE THE COARSE GRID CAPABILITY, HE MAY OMIT
71* C     ELI, AV2, AND IN2 AND CALL EXCIT DIRECTLY.
72* C
73* C
74* C                    EXAMPLE
75* C
76* C       SET UP THE COARSE GRID, PERFORM ITERATIONS ON COARSE AND FINE
77* C     GRIDS.  THEN MAKE AN ADDITIONAL N OPTIONAL PASSES THROUGH BOTH
78* C     GRIDS TO REFINE THE SOLUTION:
79* C
80* C     CALL ELI (IMAX, NX,DX, NY,DY, P,R, AX,AY, S,ERR)
81* C     DO 2 I=1,N
82* C   2 CALL ELI2(IMAX, NX,DX, NY,DY, P,R, AX,AY, S,ERR)
83* C
84* C
85*       SUBROUTINE ELI(IMAX, NX,DX,NY,DY, P, R, AX,AY, S, ERR)
86* C
87* C     SOLVE (DEL**2 + A*GRAD) P = R   AX, AY CONTAIN FACTORS 1/2DX,1/2DY
88* C     IMPROVE LONG WAVELENGTH COMPONENTS ON COARSE MESH THEN FINISH
89* C     THE SOLUTION ON THE FULL SIZE MESH.  THE HARD CORE NUMERICS ARE
90* C     DONE IN SUBROUTINE EXCIT. PERIODIC B.C.   S IS A SCRATCH ARRAY.
91* C
92*             PARAMETER MX=130, MY=130, IX=4, IY=4, MLX=(MX-2)/IX + 2,
93*      1       MLY=(MY-2)/IY + 2, MXH=(MX-2)/2 + 2, MYH=(MY-2)/2 + 2
94*             REAL P(NX/MX/,NY/MY/), R(NX/MX/,NY/MY/), AX(NX/MX/,NY/MY/),
95*      1       AY(NX/MX/,NY/MY/), S(NX/MX/,NY/MY/)
96*             COMMON/ LMESH/ PL(MLX,MLY), RL(MLX,MLY), ALX(MLX,MLY),
97*      1       ALY(MLX,MLY), S2(MXH,MYH)
98*             DATA PI/3.141593/
99* C
100* C     EXTRACT THE RESIDUAL
101*            RDX2 = 1./DX**2
102*            RDY2 = 1./DY**2
103*            NXM = NX-1
104*            NYM = NY-1
```

```
105* C
106*            DO 2 J=2,NYM
107*            DO 2 I=2,NXM
108*     2      S(I,J) = RDX2*(P(I+1,J) - 2.*P(I,J) + P(I-1,J))
109*     1             + RDY2*(P(I,J+1) - 2.*P(I,J) + P(I,J-1))
110*     1             + AX(I,J)*(P(I+1,J) - P(I-1,J))
111*     3             + AY(I,J)*(P(I,J+1) - P(I,J-1))     -    R(I,J)
112* C
113* C     REDUCE LONG WAVELENGTH ERROR ON COARSE GRID. GET AVG'S ON SMALL MESH
114*            LX = (NX-2)/IX + 2
115*            LY = (NY-2)/IY + 2
116*            NXH = (NX-2)/2 + 2
117*            NYH = (NY-2)/2 + 2
118* C
119*            DO 3 J=1,MLY
120*            DO 3 I=1,MLX
121*            PL(I,J) = 0.
122*            ALX(I,J) = 0.
123*     3      ALY(I,J) = 0.
124* C
125* C         REDUCE THE MESH BY 2 FACTORS OF 2
126*            CALL AV2(NX,NY, AX, NXH,NYH,  S2)
127*            CALL AV2(NXH,NYH, S2, LX,LY, ALX)
128*            CALL AV2(NX,NY, AY, NXH,NYH,  S2)
129*            CALL AV2(NXH,NYH, S2, LX,LY, ALY)
130* C
131*            FX = 1./IX
132*            FY = 1./IY
133*            DO 4 J=1,MLY
134*            DO 4 I=1,MLX
135*            ALX(I,J) = ALX(I,J)*FX
136*     4      ALY(I,J) = ALY(I,J)*FY
137* C
138*            ENTRY ELI2(IMAX, NX,DX, NY,DY, P, R, AX,AY, S, ERR)
139* C   THIS ENTRY ALLOWS MULTIPLE USE OF COEFFICIENTS ALX, ALY JUST DEFINED
140*            CALL AV2(NX,NY,  S, NXH,NYH,  S2)
141*            CALL AV2(NXH,NYH, S2, LX,LY,  RL)
142*            MAXL = 3*IMAX
143*            CALL EXCIT(MAXL, LX, IX*DX, LY, IY*DY, PL, RL, ALX,ALY, S, EL)
144* C
145* C         INTERPOLATE BACK ONTO LARGE MESH
146*            CALL IN2(LX,LY, PL, NXH,NYH, S2)
147*            CALL IN2(NXH,NYH, S2, NX,NY, S)
148* C
149*            DO 5 J=1,NY
150*            DO 5 I=1,NX
151*     5      P(I,J) = P(I,J) - S(I,J)
152* C      NOW CORRECT THE SHORT WAVELENGTH ERRORS
153* C      INCREASE THE MINIMUM EIGENVALUE ESTIMATE BY A FACTOR SF
154* C      THIS INCREASES SHORT WAVE CONVERGENCE BY ROUGHLY A FACTOR SF.
155* C      THE FOLLOWING APPROXIMATES RESULTS OF OPTIMUM REGRID MODEL.
156* C      SF SHOULD BE APPROX. 7 FOR 100 BY 100 MESHES.
157*            SF = .232* FLOAT( MAX0(NX-2,NY-2))**.75
```

```
158*              CALL SIGSET(SF* 2.*PI/MAX0(NX-2,NY-2) )
159*              CALL EXCIT(IMAX, NX,DX, NY,DY, P, R, AX, AY, S, ERR)
160*       RETURN
161*       END
162* C
163* C
164*       SUBROUTINE EXCIT(IMAX, NX,DX, NY,DY, P,S, AX,AY, D, ERR)
165* C     EXPLICIT CHEBYCHEV ITERATIVE SOLVER FOR
166* C     (DEL**2 + A*GRAD) P = S
167* C         ON A MESH OF CONSTANT GRID SPACING DX, DY.
168* C     RESIDUAL RETURNS IN D
169* C     USER SUPPLIES THE BOUNDARY SETTING SUBROUTINE PBCSET.
170* C     SIG IN STATEMENT 24 IS DEFINED FOR DOUBLY PERIODIC BOUNDARIES.
171* C
172* C     NOTE!!   ARRAYS AX, AY CONTAIN X & Y COMPONENTS OF VECTOR A
173* C     DIVIDED BY 2DX AND 2DY RESPECTIVELY.
174* C
175*              PARAMETER MX=130,MY=130
176*              REAL P(NX/MX/,NY/MY/), S(NX/MX/,NY/MY/), AX(NX/MX/,NY/MY/),
177*       1         AY(NX/MX/,NY/MY/), D(NX/MX/,NY/MY/)
178*              LOGICAL LPRINT /.TRUE./
179*              COMMON /SCRACH/ AXM(MX,MY), AYM(MX,MY)
180*              REAL*8 A,B,C, RAD,Q, TM,TN,TP, HMAXR
181*              REAL MU
182*              DATA KALL, IPRINT, MODPR, IENTRY, PI /0,1,1,0, 3.141593/
183*              CBRT(Y) = AMAX1(Y, 1E-30)**.333333
184* C
185*              KALL = KALL + 1
186*              RDX2 = 1./DX**2
187*              RDY2 = 1./DY**2
188*              NXM = NX-1
189*              NYM = NY-1
190* C
191* C     MAGNITUDE OF A, S
192* C
193*              DO 15 J=1,NY
194*              DO 15 I=1,NX
195*       15     D(I,J) = 0.
196*              SS = 0.
197* C
198*              DX4 = 4.*DX**4
199*              DY4 = 4.*DY**4
200*              DO 2 J=2,NYM
201*              DO 2 I=2,NXM
202*              D(I,J) = DX4*AX(I,J)**2 + DY4*AY(I,J)**2
203*       2      SS = SS + S(I,J)**2
204* C
205* C     CALL USER SUPPLIED ROUTINE TO FIND MAXIMUM ELEMENT OF ARRAY D
206*              CALL MAX2D(NX,NY, NX*NY, D, MAXI,MAXJ)
207*              ABAR = SQRT(D(MAXI, MAXJ))
208*              GO TO 24
209* C
210*       ENTRY SIGSET(SI)
```

```
211*             IENTRY = 1
212*             SIG = SI
213*          RETURN
214* C
215* C     SIG IS DETERMINED FROM THE LOWEST MODE SINUSOID CONSISTENT WITH
216* C     THE BOUNDARY CONDITIONS.
217* C     FOR DOUBLY PERIODIC B.C. :
218*    24     IF(IENTRY .EQ. 0) SIG = 2.*PI/MAXO(NX-2,NY-2)
219* C     FOR NEUMANN B.C.      SIG =   PI/MAXO(NX-2,NY-2)
220* C     FOR DIRICHLET B.C.    SIG = PI*SQRT(1./(NX-2)**2 + 1./(NY-2)**2)
221*             IENTRY = 0
222* C
223* C     DIMENSIONLESS PARAMETER FOR MEASURING IMPORTANCE OF A
224*          MU = ABAR/(4.*SIG)
225* C
226* C     CUBIC SOLUTION FOR THETA
227*          SR = SQRT(16.*MU**2 + 1.)
228*          SR1 = SR - 1.
229*          IF(SR1 .LT. 1E-4) SR1 = 8.*MU**2
230*            THETA = CBRT(.25*MU**2) * (CBRT(SR + 1.) - CBRT(SR1) )
231* C
232* C     ELLIPSE MAJOR & MINOR AXES (A,C)   &   MIDPOINT B
233*          HMAXR = DSQRT(1D0 - SIG**2*(.5 -    THETA))
234*          B = 2.*(RDX2 + RDY2)
235*          A = B*HMAXR
236*          C = A*MU*SIG/SQRT( THETA)
237*          A2 = A**2
238*          C2 = C**2
239*          IF( ABS(A2 - C2) .LT. .01*(A2 + C2)) C = A*1.025
240*          C2 = C**2
241*          SIGNAC = SIGN(1., A2 - C2)
242*          RAD = DSQRT( DABS(A**2 - C**2) )
243*          CA = C/A
244*          Q = B/RAD
245*          TM = Q*SIGNAC
246*          TN = 1.
247* C
248*          DO 28 J=1,NY
249*          DO 28 I=1,NX
250*    28    D(I,J) = 0.
251* C
252*          DO 29 J=2,NYM
253*          DO 29 I=2,NXM
254*          AXM(I,J) = RDX2 - AX(I,J)
255*          AX (I,J) = RDX2 + AX(I,J)
256*          AYM(I,J) = RDY2 - AY(I,J)
257*    29    AY (I,J) = RDY2 + AY(I,J)
258* C
259* C     ITERATION LOOP. . . . . . . . . . . . . . . . . . . . . . . . . .
260*          ITMAX = IMAX + MOD(IMAX,2)
261*          DO 4 IT=1,ITMAX
262*          TP = 2.*Q*TN - TM*SIGNAC
263* C
```

CIFER -- VERSION 06.03   DATE = 09/18/80   TIME = 13:13:52:97

```
264*              CN = 2.*TN/(RAD*TP)
265*              CM = TM/TP*SIGNAC
266*              IF(IT.EQ.1) CN = .5*CN
267* C
268* C     STORE THE ITERATE ALTERNATELY INTO P AND D.  THIS SAVES ONE STORE
269* C     AND THEREBY REDUCES OPERATION COUNT FROM 12 TO 11.
270*              IF(MOD(IT,2) .EQ. 0) GO TO 35
271*              DO 32 J=2,NYM
272*              DO 32 I=2,NXM
273*      32      D(I,J) = CN*( AX(I,J)*P(I+1,J) + AXM(I,J)*P(I-1,J)
274*       1       + AY(I,J)*P(I,J+1) + AYM(I,J)*P(I,J-1) - S(I,J)) - CM*D(I,J)
275*              CALL PBCSET(NX,NY, D)
276*              GO TO 38
277* C
278*      35      DO 36 J=2,NYM
279*              DO 36 I=2,NXM
280*      36      P(I,J) = CN*( AX(I,J)*D(I+1,J) + AXM(I,J)*D(I-1,J)
281*       1       + AY(I,J)*D(I,J+1) + AYM(I,J)*D(I,J-1) - S(I,J)) - CM*P(I,J)
282*              CALL PBCSET(NX,NY, P)
283*      38      TM = TN
284*              TN = TP
285*      4       CONTINUE
286* C     . . . . . . . . . . . . . . . . . . . . . . . . . . . . .
287* C
288* C     KILL ODD EVEN MODE & EXTRACT RESIDUAL
289*              CO = 2.*(RDX2 + RDY2)
290*              PAV = 0.
291*              DO 45 J=2,NYM
292*              DO 45 I=2,NXM
293*              PAV = PAV + P(I,J)
294*      45      D(I,J) = AX(I,J)*P(I+1,J) - CO*P(I,J) + AXM(I,J)*P(I-1,J)
295*       1       + AY(I,J)*P(I,J+1) + AYM(I,J)*P(I,J-1)  - S(I,J)
296* C
297*              PAV = PAV/((NX-2)*(NY-2))
298*              RC = .5/CO
299*              DO 5 J=2,NYM
300*              DO 5 I=2,NXM
301*      5       P(I,J) = P(I,J) + RC*D(I,J) - PAV
302*              CALL PBCSET(NX,NY, P)
303* C
304*              DO 55 J=2,NYM
305*              DO 55 I=2,NXM
306*      55      D(I,J) = AX(I,J)*P(I+1,J) - CO*P(I,J) + AXM(I,J)*P(I-1,J)
307*       1       + AY(I,J)*P(I,J+1) + AYM(I,J)*P(I,J-1)  - S(I,J)
308* C
309* C     RESET AX, AY
310* C
311*              ERR = 0.
312*              DO 6 J=2,NYM
313*              DO 6 I=2,NXM
314*              AY(I,J) = AY(I,J) - RDY2
315*              AX(I,J) = AX(I,J) - RDX2
316*      6       ERR = ERR + D(I,J)**2
```

```
317*              ERR = SQRT(ERR/SS)
318*              GO TO 62
319* C
320*          ENTRY EXPRNT(IPR, MPR)
321*          IPRINT = IPR
322*          MODPR = MPR
323*          RETURN
324* C
325*    62    LPRINT = MOD(KALL,IPRINT) .LE. MODPR
326*          IF(LPRINT) PRINT 65, NX,NY, ITMAX, MU,    CA, ERR
327*    65    FORMAT( 20X,'EXCIT NX NY ITMAX ', 3I4, '   MU   ', 1P E10.3,
328*       1            C/A ', E10.3, '    ERR ', E10.3)
329*          RETURN
330*          END
331* C
332* C
333*          SUBROUTINE AV2(NX,NY, AA, NXH,NYH, A)
334*          PARAMETER M=130, MH=66
335*          REAL AA(NX/M/,NY/M/),  A(NXH/MH/,NYH/MH/)
336* C  DEFINE ELEMENTS OF SMALL ARRAY A TO BE 4 POINT AVERAGES OF
337* C  ELEMENTS OF LARGE ARRAY AA.  THE SPATIAL ORIGIN OF A IS OFFSET
338* C  BY 1/2 ZONE IN EACH DIRECTION TO ACHIEVE THE FOLLOWING CENTERED
339* C  ARRANGEMENT OF ELEMENTS:
340* C                       AA        AA
341* C
342* C                            A
343* C
344* C                       AA        AA
345* C
346* C  2*NXH MUST = NX+2.  ETC FOR Y
347*          IF(2*NXH.EQ.NX+2 .AND. 2*NYH.EQ.NY+2) GO TO 2
348*          PRINT 15, NXH,NX, NYH,NY
349*    15    FORMAT(/// 20X, 'AV2 CALLED WITH ILLEGAL DIMENSIONS.
350*       1        'NXH NX NYH NY =', 4I10)
351*          STOP
352* C
353*     2    NXHM = NXH-1
354*          NYHM = NYH-1
355* C
356* C
357*          DO 3 JH=1,NYH
358*          DO 3 IH=1,NXH
359*     3    A(IH,JH) = 0.
360* C
361*          DO 4 JH=2,NYHM
362*          DO 4 IH=2,NXHM
363*          IM = 2*IH - 2
364*          JM = 2*JH - 2
365*     4    A(IH,JH) = A(IH,JH) + .25*( AA(IM,JM) + AA(IM+1,JM)
366*       1        + AA(IM,JM+1) + AA(IM+1,JM+1) )
367*     5    CONTINUE
368*          CALL PBCSET(NXH,NYH, A)
369*          RETURN
```

```
370* C
371*          ENTRY IN2(NXH,NYH, A, NX,NY, AA)
372* C     BILINEAR INTERPOLATION FROM SMALL MESH A ONTO LARGE MESH AA.
373*          IF(2*NXH.EQ.NX+2 .AND. 2*NYH.EQ.NY+2) GO TO 6
374*          PRINT 55, NXH,NX, NYH,NY
375*    55    FORMAT(/// 20X, 'IN2 CALLED WITH ILLEGAL DIMENSIONS.
376*     1        'NXH NX NYH NY =', 4I10)
377*       STOP
378* C
379*     6    NXHM = NXH-1
380*          NYHM = NYH-1
381* C
382*          DO 65 J=1,NY
383*          DO 65 I=1,NX
384*    65    AA(I,J) = 0.
385* C
386*          CALL PBCSET(NXH,NYH, A)
387* C
388*          DO 8 LY=1,2
389*          DO 8 LX=1,2
390*          X = .25 + .5*(LX-1)
391*          Y = .25 + .5*(LY-1)
392* C
393*          DO 8 IX=1,2
394*          DO 8 IY=1,2
395*          IF(IX.EQ.1 .AND. IY.EQ.1) F = (1.-X)*(1.-Y)
396*          IF(IX.EQ.2 .AND. IY.EQ.1) F =    X*(1.-Y)
397*          IF(IX.EQ.1 .AND. IY.EQ.2) F = (1.-X)*Y
398*          IF(IX.EQ.2 .AND. IY.EQ.2) F =    X*Y
399* C
400*          DO 7 J=1,NYHM
401*          DO 7 I=1,NXHM
402*     7    AA(2*I-2+LX,2*J-2+LY) =
403*     1    AA(2*I-2+LX,2*J-2+LY) + F*A(I+IX-1,J+IY-1)
404*     8    CONTINUE
405*       RETURN
406*       END
407* C
408* C
409*          SUBROUTINE PBC SET(NY,NZ,Q)
410* C     SET EXTERIOR GUARD CELLS OF ARRAY Q FOR DOUBLY PERIODIC B.C.
411*          REAL Q(NY,NZ)
412*          NYM = NY-1
413*          NZM = NZ-1
414* C
415*          DO 2 J=2,NYM
416*          Q(J,1) = Q(J,NZM)
417*     2    Q(J,NZ) = Q(J,2)
418* C
419*          DO 3 K=1,NZ
420*          Q(1,K) = Q(NYM,K)
421*     3    Q(NY,K) = Q(2,K)
422*       RETURN
```

```
423*        END
424* C
425* C
426*        SUBROUTINE MAX2D(NX,NY, NXY, S, IM,JM)
427*        REAL S(NXY)
428* C     THE ASC SUPPLIES VECTORIZED ROUTINES MINVAL, MAXVAL TO LOCATE
429* C     MINIMUM & MAXIMUM ELEMENTS OF SINGLY DIMENSIONED ARRAY.
430* C     TRANSLATE LOCATIONS AS APPROPRIATE FOR DOUBLY DIMENSIONED S(NX,NY)
431*        NM = MAXVAL(S)
432*        GO TO 2
433*        ENTRY MIN2D(NX,NY, NXY, S, IM,JM)
434*        NM = MINVAL(S)
435*    2   JM = 1 + NM/NX
436*        IM = 1 + NM - (JM-1)*NX
437*        RETURN
438*        END
```

References

Ablowitz, M. J., D. J. Kaup, A. C. Newell, and H. Segur, The Inverse Scattering Trans-
form-Fourier Analysis for Nonlinear Problems, Stud. Appl. Math. **53**, 249 (1974).

Ben-Dor, G., and I. I. Glass, Nonstationary Oblique Shock Reflection: Actual Isopcnics
and Numerical Examples, AIAA J. **16**, 1146 (1978).

Benjamin, T. B., Internal Waves of Permanent Form in Fluids of Great Depth, J. Fluid
Mech. **29**, 559 (1967).

Birdsall, C. K., and D. Fuss, Clouds-in-Clouds, Clouds-in-Cells Physics for Many-Body
Plasma Simulation, J. Comp. Phys. **3**, 494 (1969).

Book, D. L., R. L. Burton, A. L. Cooper, R. D. Ford, B. Hui, D. J. Jenkins, P. C. Liewer,
A. E. Robson, P. J. Turchi, S. Hamasaki, N. A. Krall, L. Mascheroni, R. Shanny,
E. L. Cantrell, R. A. Gerwin, I. Henins, R. W. Moser, G. A. Sawyer, A. R. Sherwood,
B. R. Suydan, and C. E. Swannak, Experimental and Theoretical Liner Fusion
Studies, *Plasma Physics and Controlled Nuclear Fusion Research 1978* (IAEA,
Vienna, 1979), Vol. 2, p. 93.

Book, D. L., J. P. Boris, and K. H. Hain, Flux-Corrected Transport II: Generalizations
of the Method, J. Comp. Phys. **18**, 248 (1975).

Book, D. L., J. P. Boris, A. L. Kuhl, E. S. Oran, and S. T. Zalesak, Simulation of Complex
Shock Reflection from Wedges in Inert and Reactive Gaseous Mixtures, Seventh
International Conf. on Numerical Methods in Fluid Dynamics, Stanford, CN, June
23–27, 1980.

Book, D. L., and P. J. Turchi, Dynamics of Rotationally Stabilized Implosions of Compressible Cylindrical Liquid Shells, Phys. Fluids **22**, 68 (1979).

Boris, J. P., Flux-Corrected Transport Modules for Generalized Continuity Equations,
NRL Memo. Rept. 3237 (1976).

Boris, J. P., Numerical Solution of Continuity Equations, *Proc. Second European Conference on Computational Physics* (North-Holland, Amsterdam 1976a); NRL Memo.
Rept. 3327.

Boris, J. P., ADINC: An Implicit Lagrangian Hydrodynamics Code, NRL Memo. Rept.
4022 (1979).

Boris, J. P., and D. L. Book, Flux-Corrected Transport I: SHASTA—A Fluid Transport
Algorithm That Works, J. Comp. Phys. **11**, 38 (1973).

Boris, J. P., and D. L. Book, Flux-Corrected Transport III: Minimal-Error FCT Algorithms, J. Comp. Phys. **20**, 397 (1976).

Boris, J. P., and D. L. Book, Solution of Continuity Equations by the Method of Flux-
Corrected Transport, *Methods in Computational Physics* (Academic Press, New
York, 1976a), Vol. 16, p. 85.

Boris, J. P., M. J. Fritts, and K. L. Hain, Free Surface Hydrodynamics Using a Lagran-gain Triangular Mesh, in *Proc. First Int'l. Conference on Numerical Ship Hydrody-namics* (NBS, Gaithersburg, Md., 1975).

Boris, J. P., and K. V. Roberts, The Optimization of Particle Calculations in 2 and 3 Dimensions, J. Comp. Phys. **4**, 552 (1969).

Boris, J. P., and R. A. Shanny, Eds., *Proc. Fourth Conference on Numerical Simulation of Plasmas* (U.S. GPO, Washington, D.C., 1970).

Brennan, C., and A. K. Whitney, Unsteady Free Surface Flows; Solutions Employing the Lagrangian Description of the Motion, Eighth Symposium on Naval Hydro-dynamics, Pasadena, CA, August, 1970.

Buneman, O., Time-Reversible Difference Procedures, J. Comp. Phys. **1**, 517 (1967).

Buneman, O., A Compact Non-iterative Poisson Solver, SUIPR Report No. 294, Inst. for Plasma Physics, Stanford University, 1969.

Burridge, D. M., A Split Semi-Implicit Reformulation of the Bushby–Thompson 10-Level Model, Q. J. Roy. Met. Soc. **101**, 777 (1975).

Chan, R. K.-C., A Generalized Arbitrary Lagrangian-Eulerian Method for Incompress-ible Flows with Sharp Interfaces, J. Comp. Phys. **17**, 311 (1974). See also C. W. Hirt, A. A. Amsden, and J. L. Cook, "An Arbitrary Lagrangian-Eulerian Computing Method for all Flow Speeds," J. Comp. Phys. **14**, 227 (1974).

Chandrasekhar, S., *Hydrodynamics and Hydromagnetic Stability* (Clarendon Press, London, 1961), p. 428.

Christiansen, J. P., Numerical Simulation of Hydrodynamics by the Method of Point Vortices, J. Comp. Phys. **13**, 363 (1973).

Cooley, J. W., and Tukey, J. W., An Algorithm for the Machine Computation of Com-plex Fourier Series, Math. Comp. **19**, 297 (1965).

Courant, R., and K. O. Friedrichs, *Supersonic Flow and Shock Waves* (Interscience, New York, 1948).

Crout, P. D., A Short Method for Evaluating Determinants and Solving Systems of Linear Equations with Real or Complex Coefficients, Trans. AIEE **60**, 1235 (1941).

Crowley, W. P., A Global Numerical Ocean Model: Part I, J. Comp. Phys. **3**, 111 (1968).

Crowley, W. P., Flag, A Free-Lagrange Method for Numerically Simulating Hydro-dynamics Flows in Two Dimensions, *Proc. Second International Conference on Numerical Methods in Fluid Dynamics* (Springer-Verlag, New York, 1971).

Dietrich, D., B. E. McDonald, and A. Warn-Varnas, Optimized Block-Implicit Relax-ation, J. Comp. Phys. **18**, 421 (1975).

Evans, M. W., and F. H. Harlow, The Particle-in-Cell Method for Hydrodynamic Calculations, Los Alamos Scientific Laboratory Report No. LA-2139 (1957).

Forester, C. K., Higher Order Monotonic Convective Difference Schemes, J. Comp. Phys. **23**, 1 (1977).

Fornberg, B., On a Fourier Method for the Integration of Hyperbolic Equations, SIAM J. Num. Anal. **12**, 509 (1975).

Fritts, M. J., A Numerical Study of Free-Surface Waves, SAI Report SAI-76-528-WA (1976).

Fritts, M. J., Lagrangian Simulations of the Kelvin-Helmholtz Instability, SAI-76-632-WA (1976a).

Fritts, M. J., and J. P. Boris, Solution of Transient Problems in Free-Surface Hydro-dynamics, J. Comp. Phys., **31**, 173 (1979).

Fritts, M. J., F. W. Miner, and O. M. Griffin, Numerical Calculation of Wave-Structure Interactions, *Computer Methods in Fluids* (Pentech Press, 1980), Chap. 1.

Gadd, A. J., A Split-Explicit Integration Scheme for Numerical Weather Prediction, Q. J. Roy. Met. Soc. **104**, 569 (1978).

Gelinas, R. J., S. K. Doss, and K. Miller, The Finite-Element Method with Moving Nodes: A Computational Physics Breakthrough, Science Applications Report SAI/PL/F179 (1979).

Gottlieb, D., Strang-Type Difference Schemes for Multidimensional Problems, SIAM J. Num. Anal. **9**, 650 (1972).

Gottlieb, D., and S. A. Orszag, *Numerical Analysis of Spectral Methods: Theory and Applications* (SIAM, Philadelphia, 1977).

Gottlieb, D., and S. A. Orszag, Spectral Calculations of One-Dimensional Inviscid Compressible Flows, Cambridge Hydrodynamics Report No. 27 (1979).

Hain, K., The Partial Donor Cell Method, NRL Memo. Rept. 3713 (1978).

Hamming, R. W., *Numerical Methods for Scientists and Engineers* (McGraw-Hill, New York, 1962).

Harlow, F. H., The Particle-in-Cell Computing Method for Fluid Dynamics, *Methods in Computational Physics*, B. Alder, S. Fernbach, and M. Rotenberg, Eds. (Academic Press, New York, 1964), Vol. 3, p. 319.

Harlow, F. H., and J. F. Welch, Numerical Calculation of Time-Dependent Viscous Incompressible Flow of Fluid with Free Surface, Phys. Fluids **18**, 2182 (1965).

Harten, A., The Method of Artificial Compression, CIMS Rept. COO-3077-50 (1974).

Hirota, I., T. Tokioka, and M. Nishiguchi, A Direct Solution of Poisson's Equation by Generalized Sweep-out Method, J. Meteor. Soc. Japan **48**, 161 (1970).

Hockney, R. W., A Fast Direct Solution of Poisson's Equation Using Fourier Analysis, Comm. Assoc. Comp. Mach. **12**, 95 (1965).

Hockney, R. W., "Further Computer Experimentation on Anomalous Diffusion," SUIPR Rept. No. 202, Inst. for Plasma Research, Stanford Univ., Stanford, Calif. (1966).

Hockney, R. W., The Potential Calculation and Some Applications, *Methods in Computational Physics*, B. Alder, S. Fernback, and M. Rotenberg, Eds. (Academic Press, New York, 1970), Vol. 9.

Hoskin, N. E., Solution by Characteristics of the Equations of One-Dimensional Unsteady Flow, *Methods in Computational Physics*, B. Alder, S. Feinbach, and M. Rotenberg, Eds. (Academic Press, New York, 1964), Vol. 3, p. 265.

Jones, W. W., and J. P. Boris, Flame and Reactive Jet Studies Using a Self-Consistent Two-Dimensional Hydrocode, J. Phys. Chem. **81**, 2532 (1977).

Kershaw, D., The Incomplete-Cholesky–Conjugate-Gradient Method for the Iterative Solution of Systems of Linear Equations, J. Comp. Phys. **26**, 43 (1978).

Kreiss, H. O., A Comparison of Numerical Methods Used in Atmospheric and Oceanographic Applications, in Proc. of the Symposium on Numerical Models of Ocean Circulation, National Academy of Sciences (U.S. Government Printing Office, Washington, D.C., 1972). See also in the same volume B. Wendroff, Problems of Accuracy with Conventional Finite-Difference Methods, and G. Fix, A Survey of Numerical Methods for Selected Problems in Continuum Mechanics.

Kreiss, H. O., and J. Oliger, Comparison of Accurate Methods for the Integration of Hyperbolic Equations, Tellus **24**, 199 (1972).

Kulikovsky, A. G., and G. A. Lyubimov, *Magnetohydrodynamics* (Addison–Wesley, Reading, MA, 1965).

Lanczos, C., *Applied Analysis* (Prentice–Hall, Englewood Cliffs, N.J., 1956).

Landau, L. D., and E. M. Lifshitz, *Fluid Mechanics* (Addison–Wesley, Reading, MA, 1959).

Langdon, A. B., and B. J. Lasinski, Electromagnetic and Relativistic Plasma Simulation Models, *Methods in Computational Physics* (Academic Press, New York, 1976), Vol. 16, p. 327.

Launder, B. E., and D. B. Spalding, *Mathematical Models of Turbulence* (Academic Press, London, 1972).

Lax, P. D., Periodic Solutions of the KdV Equation, Comm. Pure Appl. Math. **28**, 141 (1975).

Lindzen, R. S., and H.-L. Kuo, A Reliable Method for the Numerical Integration of a Large Class of Ordinary and Partial Differential Equations, Mon. Weath. Rev. **97**, 732 (1969).

Madala, R. V., An Efficient Direct Solver for Separable and Non-Separable Elliptic Equations, Mon. Weath. Rev. **106**, 1735 (1978).

Madala, R. V., A Limited Area Numerical Model Suitable for Numerical Weather Prediction, in press.

Madala, R. V., and S. A. Piacsek, A Semi-Implicit Numerical Model for Baroclinic Ocean, J. Comp. Phys. **23**, 167 (1977).

Manteuffel, T. A., The Tchebychev Iteration for Nonsymmetric Linear Systems, Numer. Math. **28**, 307 (1977).

Marder, B. M., GAP—A PIC-Type Fluid Code, Math. Comp. **29**, 434 (1975).

Marchuk, G. I., *Numerical Methods in Weather Prediction* (Academic Press, New York, 1974).

McAvaney, B. J., and L. M. Leslie, Comments on a Direct Solution of Poisson's Equation by Generalized Sweep-out Method, J. Meteor. Soc. Japan **50**, 136 (1972).

McDonald, B. E., The Chebychev Method for Solving Non-Self-Adjoint Elliptic Equations on a Vector Computer, J. Comp. Phys. **35**, 147 (1980).

McDonald, B. E., T. P. Coffey, S. L. Ossakow, and R. N. Sudan, Numerical Studies of Type 2 Equatorial Electrojet Irregularity Development, Radio Sci. **10**, 247 (1975).

McDonald, B. E., S. L. Ossakow, S. T. Zalesak, and N. J. Zabusky, A Fluid Model for Estimating Minimum Scale Sizes in Ionospheric Plasma Cloud Striations, NRL Memo. Rept. 3864 (1978).

Miner, E. W., O. M. Griffin, S. E. Ramberg, and M. J. Fritts, Numerical Calculation of Wave Effects on Structures, *Proc. Civil Engineering in the Oceans IV* San Francisco, Ca., 10–12 Sept., 1979.

Miura, R. M., The Korteweg–de Vries Equation: A Model Equation for Nonlinear Dispersive Waves, *Nonlinear Waves*, S. Leibowitz and A. R. Seebass, Eds. (Cornell Univ. Press, Cornell, NY, 1974).

Moretti, G., and M. Abbott, A Time-Dependent Computational Method for Blunt Body Flows, AIAA J. **4** (No. 12) 1966.

Morse, R. L., Multidimensional Plasma Simulation by the Particle-in-Cell Methods, *Methods in Computational Physics*, B. Alder, S. Fernbach, and M. Rotenberg, Eds. (Academic Press, New York, 1970), Vol. 9.

O'Brien, J. J., and H. F. Hurlburt, A Numerical Model of Coastal Upwelling, J. Phys. Ocean. **2**, 14 (1972).

Oran, E. S., and J. P. Boris, Theoretical and Computational Approach to Modelling Flame Ignition, NRL Memo. Rept. No. 4131 (1979).

Oran, E. S., T. R. Young, Jr., and J. P. Boris, Application of Time-Dependent Numerical Methods to the Description of Reactive Shocks, *Proc. 17th Symposium (International on Combustion)*, (Combustion Institute, Pittsburgh, PA, 1980), p. 43.

Orszag, S. A., Spectral Methods for Problems in Complex Geometries, *Advances in Computer Methods for Partial Differential Equations—III*, R. Vichnevetsky and R. S. Stepleman, Eds. (IMACS, Rutgers Univ., New Brunswick, NJ, 1979), p. 148.

Orszag, S. A., and M. Israeli, Numerical Flow Simulation by Spectral Methods, *Proc. of the Symposium on Numerical Models of Ocean Circulation*, (U.S. Government Printing Office, Washington, D.C., 1972).

Ott, E., W. M. Manheimer, D. L. Book, and J. P. Boris, Model Equations for Mode Coupling Saturation in Unstable Plasmas, Phys. Fluids **16**, 855 (1973).

Richtmyer, R. D., and K. W. Morton, *Difference Methods for Initial Value Problems* (Interscience, New York, 1967).

Roache, P. J., A New Direct Method for the Discretized Poisson Equation, *Proceedings of the Second International Conference on Numerical Methods in Fluid Dynamics* (Springer-Verlag, New York, 1971).

Roache, P. J., *Computational Fluid Dynamics* (Hermosa Publishers, Albuquerque, 1975).

Roache, P. J., Marching Methods for Elliptic Problems: Part I, Numer. Heat Transfer **1**, 1 (1978); Part II, ibid. **1**, 163 (1978); Part III, ibid. **1**, 198 (1978).

Robert, A., J. Henderson, and C. Turnbull, A Implicit Time Interpretation Scheme for Baroclinic Models of the Atmosphere, Mon. Weath. Rev. **100**, 329 (1972).

Rosmond, T. E., and F. D. Faulkner, Direct Solution of Elliptic Equations by Block Cyclic Reduction and Factorization, Mon. Weath. Rev. **104**, 641 (1976).

Scannapieco, A. J., S. L. Ossakow, S. R. Goldman, and J. M. Pierre, Plasma Cloud Late Time Striation Spectra, J. Geophys. Res. **81**, 6037 (1976).

Schoeberl, M. R., Direct Elliptic Equation Solvers with Low Memory Requirements, NRL Memo Rept. 4191 (1980).

Scott, A. C., F. Y. F. Chu, and D. McLaughlin, The Soliton: A New Concept in Science, Proc. IEEE. **61**, 1443 (1973).

Steinberg, D., R. E. Kidder, and A. B. Cecil, A One-Dimensional Magnetohydrodynamics Code, Lawrence Radiation Laboratory Report UCRL-14931 (1966).

Strang, G., and G. Fix, *An Analysis of the Finite Element Method* (Prentice-Hall, Engle Wood Cliffs, NJ, 1973).

Taggart, K. A., R. L. Morse, R. L. McCrory, and R. N. Remund, Two Dimensional Calculations of Asymmetric Laser Fusion Targets Using IRIS, Bull. Am. Phys. Soc. **20**, 1378 (1975).

Tappert, F., *Numerical Solutions of the KdV Equation and Its Generalizations by the Split-Step Fourier Method*, Lect. Appl. Math. **15**, (Am. Math. Soc., Providence, RI, 1974).

Van Leer, B., Toward the Ultimate Conservative Difference Scheme, J. Comp. Phys. **14**, 361 (1974).

Van Leer, B., Toward the Ultimate Conservative Difference Scheme: III. Upstream Centered Finite-Difference Schemes for Ideal Conservative Flow, J. Comp. Phys. **23**, 263 (1977).

Van Leer, B., Toward the Ultimate Conservative Difference Scheme: IV. A New Approach to Numerical Convection, J. Comp. Phys. **23**, 276 (1977a).

Varga, R. S., *Matrix Iterative Analysis* (Prentice-Hall, Englewood Cliffs, NJ, 1962).

Wilhelmson, R. B., and J. H. Ericksen, Direct Solution of Poisson's Equation in Three Dimensions, J. Comp. Phys. **25**, 319 (1977).

Winsor, N. K., and J. M. Pierre, Vectorizable Sparse Matrix Solution Methods, Bull. Am. Phys. Soc. **23**, 898 (1978).

Wurtele, M. G., On the Problem of Truncation Error, Tellus **13**, 379 (1961).

Young, T. R., Jr., and J. P. Boris, A Numerical Technique for Solving Stiff Ordinary Differential Equations Associated with the Chemical Kinetics of Reactive Flows, J. Phys. Chem. **81**, 2424 (1977).

Zabusky, N. J., A Synergetic Approach to Problems of Nonlinear Dispersive Wave Propagation and Interaction, *Proc. Symp. on Nonlinear Partial Differential Equations* (Academic Press, New York, 1967).

Zabusky, N. J., and M. D. Kruskal, Interaction of "Solitons" in a Collisionless Plasma and the Recurrence of Initial States, Phys. Rev. Lett. **15**, 240 (1965).

Zalesak, S. T., Fully Multidimensional Flux-Corrected Transport Algorithms for Fluids, J. Comp. Phys. **31**, 335 (1979).

Index

Springer Series
in Computational Physics

Editors: W. Beiglböck, H. Cabannes, H. B. Keller, J. Killeen, S. A. Orszag